MW00984912

LOGIC
magazine

9

nature
winter 2019

LOGIC

ISSUE 9: Nature

EDITORIAL

Ben Tarnoff, Moira Weigel, Jen Kagan, Alex Blasdel

CREATIVE

Xiaowei R. Wang, Celine Nguyen

PRODUCTION

Jim Fingal, Christa Hartsock

/*

Cover: Celine Nguyen

Interstitials: Xiaowei R. Wang

Images: p. 38 Google Patents, p. 120 Alex Davies, p. 123, 125 courtesy of artist, p. 169 courtesy of Mafalda Paiva, Parmenides Ltd., and the University of Atacama.

ISBN 978-0-9986626-9-5

ISSN 2573-4504

ELECTRONIC TELEGRAM

editors@logicmag.io

INTERNET

https://logicmag.io

Logic Magazine is published by the Logic Foundation, a California nonprofit public benefit corporation organized for the specific purpose of promoting education about technology.

*/

nature

LOGIC
MAGAZINE

Editorial

Chatlogs

Features

Patches

Assets

Contributors

Alyssa Battistoni is a political theorist and postdoctoral fellow at Harvard University. She is the coauthor of *A Planet to Win: Why We Need a Green New Deal.*

Tega Brain is an Australian-born artist and environmental engineer. She is an assistant professor of Digital Media at New York University.

Nick Estes is a citizen of the Lower Brule Sioux Tribe, assistant professor of American Studies at the University of New Mexico, and cofounder of The Red Nation.

Ellyn Gaydos is a farmer in New York and is writing a book of stories on the nature of seasonal work.

Donna Haraway is a Distinguished Professor Emerita in the History of Consciousness Department at the University of California, Santa Cruz.

Os Keyes is a researcher and essayist based at the University of Washington.

Anne Pasek is a postdoctoral fellow in Transitions in Energy, Culture, and Society at the University of Alberta.

Tiare Ribeaux is a new media and interdisciplinary Hawaiian-American artist and curator based in the Bay Area.

Thea Riofrancos is an assistant professor of political science at Providence College. She is the coauthor of *A Planet to Win: Why We Need a Green New Deal*.

Sara Stoudt is a PhD candidate in statistics at the University of California, Berkeley and a Berkeley Institute for Data Science Fellow.

Ben Tarnoff is a founding editor of *Logic*.

The Last Man

1/

The narratologists, and screenwriters, say that there are only three stories: man against man, man against society, man against nature. But the geologists say there is just one: We have now entered an epoch when human activity is the main force shaping planet Earth. They can see it in the sediment. Even the rocks are not safe.

This means that nature, if by nature you mean somewhere out of human reach, no longer exists. To say that there is no nature outside of culture used to sound like a poststructuralist parlor game. Now it is common sense.

If human society has become a natural force, then nature is social. More and more struggles among men, otherwise known as politics, will concern it directly. Many struggles already do. Droughts and storms send refugees fleeing across borders. Countries race to stake out their slice of a melting Arctic.

"Welcome to the Anthropocene!" the cover of *The Economist* announced several years ago, as if Earth were an airport.

The internet may be the only place high enough to get a bird's-eye view of the unfolding catastrophe. In one tab, the Amazon is burning. In another, Greenland is melting *and* burning. In the fall, the hurricanes hit. In the fall, when the fires start, the homeless come coughing into the emergency rooms. They will never be able to pay the bill. They are in good company: the richest men on the planet cannot pay for what they have taken.

Is it in "our" human nature to act this way?

We like to think not.

Meanwhile, a brave girl in a borrowed boat sets out across the Atlantic.

2/

We have been told two things about the relationship between technology and nature. The first is that technology has enabled humans to master nature. The second is that technology has caused humans to destroy nature.

At the intersection of these two stories lies the idea of the Anthropocene. The Anthropocene contains a paradox: the term recognizes the immense power humans wield over the rest of creation, such that nothing on the planet is immune. Yet this same power poses a serious threat to humans. We've shaped the earth so intensively to suit our needs that it can no longer support them. (Some of us. Some needs.)

In this issue, we try to tell a different story about the entanglements of nature and technology. No surprise that the end of the world looms large. Big Tech teams up with Big Oil to build systems for smarter drilling. The residents of a small town continue to fall ill long after the microchip plant shuts down.

But there are also reasons for optimism. There are movements demanding a more "correct relation with the nonhuman world," to borrow one contributor's phrase. This issue offers some materials for imagining what such a relation might look like.

3/

Technological mastery is a myth. Prometheus is not coming. In truth, everything is dirty, even the digital—especially the digital. Computers were supposed to be made of sunshine: "all light and clean because they are nothing but signals," as another contributor to this issue famously wrote on her first computer, an HP-86, decades ago. As she already knew then, they are less pristine than promised. Their metaphors are ethereal but their footprints are filthy. They too are implicated in armageddon.

The renewable transition itself may involve new kinds of destruction.

But recognizing that nature is human-entangled and vice versa opens up more options than conservation. Recognizing that there was never any Eden to return to lets you look ahead. Indeed, the most hopeful futures may come from the darkest histories, where the lessons of resistance have been well learned. The world has ended before; there have been many armageddons. But this also means: we have to learn how to mourn. To mourn without despair; to mourn towards a future. ⌇⌇⌇

Oil is The New Data

by Zero Cool

Big Tech is forging a lucrative partnership with Big Oil, building a new carbon cloud that just might kill us all.

I remember being nervous when I flew into Atyrau, Kazakhstan. Before boarding the flight, one of the business managers who organized the trip sent me a message with precise instructions on how to navigate the local airport:

> *Once you land, get into the bus on the right side of the driver. This side opens to the terminal. Pass through immigration, pick up your luggage, and go through Customs. The flight crew will pass out white migration cards. Fill them out, and keep it with your passport. You will need to carry these documents on you at all times once you've landed.*

Another coworker, who had flown in the night before, warned us not to worry if we found ourselves in jail. *Don't panic if you find yourself in jail. Give me a call and we'll bail you out.* Maybe she was joking.

The flight itself was uncanny. I was flying in from Frankfurt, but it felt a lot like a local American flight to somewhere in the Midwest. The plane was filled with middle-aged American businessmen equipped with black Lenovo laptops and baseball caps. The man next to me wore a cowboy-esque leather jacket over a blue-collared business shirt.

After I landed in Atyrau's single-gate airport, I located my driver, who was holding a card with my name on it. He swiftly led me into a seven-seater Mercedes van and drove me to my hotel, one of the only hotels in the city. Everyone from the flight also seemed to stay there. The drive was short. The city was overwhelmingly gray. Most of it was visibly poor. The hotel was an oasis of wealth.

Across from the hotel was another one of these oases: a gated community with beige bungalows. This was presumably where the expats who worked for Chevron lived. There was a Burger King and a KFC within walking distance. Everyone spoke a bit of English.

Security was taken extremely seriously. Each time we entered one of Chevron's offices, our passports were checked, our bags were inspected, and our bodies were patted down. Video cameras were mounted on the ceilings of the hallways and conference rooms. We were instructed to travel only using Chevron's fleet of taxis, which were wired up with cameras and mics.

All of this — Atyrau's extreme security measures and the steady flow of American businesspeople — comes from the fact that the city is home to Kazakhstan's biggest and most important oil extraction project. In 1993, shortly after the fall of the Soviet Union, the newly independent nation opened its borders to foreign investment. Kazakhstan's state-owned energy company agreed to partner with Chevron in a joint venture to extract oil.

The project was named Tengizchevroil, or TCO for short, and it was granted an exclusive forty-year right to the Tengiz oil field near Atyrau. Tengiz carries roughly 26 billion barrels of oil, making it one of the largest fields in the world. Chevron has poured money into the joint venture with the goal of using new technology to increase oil production at the site. And I, a Microsoft engineer, was sent there to help.

> "*Big Tech and Big Oil are closely linked, and only getting closer.*"

Cloud Wars

Despite the climate crisis that our planet faces, Big Oil is doubling down on fossil fuels. At over 30 billion barrels of crude oil a year, production has never been higher. Now, with the help of tech companies like Microsoft, oil companies are using cutting-edge technology to produce even more.

The collaboration between Big Tech and Big Oil might seem counterintuitive. Culturally, who could be further apart? Moreover, many tech companies portray themselves as leaders in corporate sustainability. They try to out-do each other in their support for green initiatives. But in reality, Big Tech and Big Oil are closely linked, and only getting closer.

The foundation of their partnership is the cloud. Cloud computing, like many of today's online subscription services, is a way for companies to rent servers, as opposed to purchasing them. (This model is more specifically called the *public cloud*.) It's like choosing to rent a movie on iTunes for $2.99 instead of buying the DVD for $14.99. In the old days, a company would have to run its website from a server that it bought and maintained itself. By using the cloud, that same company can outsource its infrastructure needs to a cloud provider.

"For Amazon, Google, and Microsoft, as well as a few smaller cloud competitors like Oracle and IBM, winning the IT spend of the Fortune 500 is where most of the money in the public cloud market will be made."

The market is dominated by Amazon's cloud computing wing, Amazon Web Services (AWS), which now makes up more than half of all of Amazon's operating income. AWS has grown fast: in 2014, its revenue was $4.6 billion; in 2019, it is set to surpass $36 billion. So many companies run on AWS that when one of its most popular services went down briefly in 2017, it felt like the entire internet stopped working.

Joining the cloud business late, Google and Microsoft are now playing catch-up. As cloud computing becomes widely adopted, Amazon's competitors are doing whatever they can to grab market share. Over the past several years, Microsoft has reorganized its internal operations to prioritize its cloud business. It is now spending tens of billions of dollars every year on constructing new data centers around the planet. Meanwhile, Google CEO Sundar Pichai announced that in 2019, the company is putting

$13 billion into constructing new offices and data centers in the US alone, the majority of which will go to the latter.

Startups have long been the biggest early adopters of the public cloud. They are an obvious fit: they do not own their own data centers, so the opportunity cost of switching to the public cloud is low. By contrast, it is much harder for large companies that do run their own data centers to make the leap, since it would require selling or retiring those centers.

This helps explain why cloud providers have only captured about 30 percent of the total addressable market. While cloud technology has matured considerably over the past half-decade, big corporations that run their own data centers still dominate the majority of the world's IT infrastructure. For Amazon, Google, and Microsoft, as well as a few smaller cloud competitors like Oracle and IBM, winning the IT spend of the Fortune 500 is where most of the money in the public cloud market will be made. And among those large companies, Big Oil sits at the top. Out of the biggest ten companies in the world by revenue, six are in the business of oil production. In order words, the success of Big Oil, and the production of fossil fuels, are key to winning the cloud race.

Making Friends

In 2017, Chevron signed a seven-year deal with Microsoft, potentially worth hundreds of millions of dollars, to establish Microsoft as its primary cloud provider. Oil companies like Chevron are the perfect customer for cloud providers. For years, they have been generating enormous amounts of data about their oil wells. Chevron alone has thousands of oil wells around the world, and each well is covered with sensors that generate more than a terabyte of data per day. (A terabyte is 1,000 gigabytes.)

"In recent years, Big Tech has aggressively marketed the transformative potential of the public cloud and AI/ML to Big Oil, with great success."

At best, Chevron has only been able to use a fraction of that data. One problem is the scale of computation required. Many servers are needed to perform the complex workloads capable of analyzing all of this data. As a result, computational needs may skyrocket—but then abruptly subside when the analysis is complete. These sharp fluctuations can put significant pressure on a company like Chevron. During spikes, their data centers lack capacity. During troughs, they sit idly.

This is where the promise of the public cloud comes in. Oil companies can solve their computational woes by turning to the cloud's renting model, which gives them as many servers as they need and allows them to pay only for what they use.

But Big Tech doesn't just supply the infrastructure that enables oil companies to crunch their data. It also offers many of the analytical tools themselves. Cloud services provided by Microsoft, Amazon, and Google have the ability to process and analyze vast amounts of data. The tech giants are also leaders in artificial intelligence and machine learning (AI/ML), a field focused on teaching computer systems to automatically perform complex tasks by "learning" from data. With AI/ML, oil companies can make better sense of all the data they are collecting, and can discover patterns that may help them make their operations more efficient and less costly.

AI/ML gives Big Oil yet another reason to depend on Big Tech: the level of sophistication often requires delving into the cutting edge of a field that the tech titans dominate. And if sharing their AI/ML expertise means getting a leg up on the competition in the cloud market, tech companies are more than willing to help.

In recent years, Big Tech has aggressively marketed the transformative potential of the public cloud and AI/ML to Big Oil, with great success. In 2017, Microsoft signed its seven-year contract with Chevron; in 2018, it announced major partnerships with oil giants BP and Equinor; and in 2019, it signed a deal with ExxonMobil that Exxon claims is "the industry's largest [contract] in cloud computing." Amazon recently opened an AWS office in Houston, the US oil and gas hub, and has been hiring AI/ML experts specifically to work on fossil fuel projects. Google has also developed deep relationships in the industry, partnering with Total, Anadarko Petroleum, and Nine Energy, and appointing Darryl Willis, an oil veteran, to lead Google Cloud's newly formed Oil, Gas & Energy division. Whatever the tech giants are telling their friends in the fossil fuel industry, it's working.

Drill Baby Drill

The multi-million-dollar partnership between Microsoft and Chevron was the reason I went to Kazakhstan. Microsoft sent me to Atyrau for a week-long workshop to help the Tengiz oil field adopt our technology. I was there to talk about computer vision, a field of AI/ML that gives computers the ability to understand digital images, but the workshop covered a range of topics in both AI/ML and cloud computing. We held it for a team at TCO tasked with boosting daily oil production from 600,000 barrels to 1 million. They wanted to learn about how Microsoft technology could help them modernize their oil field and increase efficiency.

The workshop took place in a large conference room in one of the TCO office buildings. The building itself wasn't particularly fancy. The exterior was run-down: it looked like it was last renovated in the 1980s. Aside from the security guards dressed in dark clothing, the interior was mostly white, with bright marble floors. The only bits of color came from the biscuits and pastries that were laid out on tables in front of the conference rooms.

At the workshop, I gave a short technical demonstration about running computer vision at scale on Microsoft's cloud computing platform. There were about forty people in the audience, predominantly businesspeople. My presentation felt like a marketing technique: the point was to flex Microsoft's engineering prowess to a technically illiterate business crowd. I made sure to include a lot of engineering jargon: "distributed training," "offline scoring," "Docker-compatible."

On the third day of the workshop, a small group of us convened at TCO headquarters in Atyrau to discuss specific AI/ML scenarios they wanted to implement. The meeting room was much nicer than where the workshop was held. It featured new videoconferencing equipment and plush ergonomic chairs. A half-dozen TCO managers were present. Yet, strangely, none of their technical staff attended. The TCO managers were mostly Americans and, with one exception, all white men. They wore monochrome suits and polished leather shoes. I felt out of place wearing sneakers and an oversized button-down. There was not a single Kazakhstani in the room.

To kick off the meeting, a Microsoft account manager gave a PowerPoint presentation that discussed common problems in the oil and gas industry that could be solved using AI/ML. One of the most complex use-cases involved using AI/ML to improve oil exploration. The traditional way to find a new oil or gas deposit is to perform a seismic survey. This is a technique that sends sound waves into the earth and then analyzes the time it takes for those waves to reflect off of different geological features. Because the data is volumetric and spans hundreds of kilometers at a minute granularity, the data collected from a single seismic survey can run over a petabyte. (A petabyte is a million gigabytes.) The output of this data is a 3D geological map, which geophysicists can study in order to recommend promising locations to build wells.

However, interpreting this map is a long and labor-intensive process. It can take months and involve many geophysicists. To make the process more efficient, computer vision technology can automatically segment different geological features to help geophysicists understand the 3D data and identify where best to drill. It seemed like a perfect example of the partnership I had been sent to Kazakhstan to help forge: a technically sophisticated and computationally intensive undertaking that played to the strengths of Big Tech while advancing a core priority of Big Oil, which was to dig more fossil fuels out of the ground while cutting costs.

Big Oil Is Watching

But the TCO managers also wanted to talk about something else. "We have a lot of workers in the oil fields. It would be nice to know where they are and what they are doing," one manager said. "If they are doing anything at all."

This is what our Chevron partners were most keen to discuss: how to better surveil their workers. TCO had thirty or forty thousand workers on site, nearly all local Kazakhstanis. They worked on rotating shifts—twelve-hour days for two weeks at a time—to keep the oil field running around the clock. And the managers wanted to use AI/ML to keep a closer eye on them.

They proposed using AI/ML to analyze the video streams from existing CCTV cameras to monitor workers throughout the oil field. In particular, they wanted to implement computer vision algorithms that could detect suspicious activity and then identify the worker engaging in that activity. (My Microsoft colleagues and I doubted the technical feasibility of this idea.) Enhancing workplace safety would be the reason for building this system, the managers claimed: more specifically, they hoped to see whether workers were drunk on site so that they could dispatch help and prevent them from hurting themselves. But in order to implement this safety measure, an "always-on" algorithmic monitoring system would have to be put in place — one that would also happen to give management a way to see whether workers were slacking off.

"Did I really want to help Chevron destroy the planet?"

The TCO managers also talked about using the data from the GPS trackers that were installed on all of the trucks used to transport equipment to the oil sites. They told us that the workers were not trustworthy. Drivers would purportedly steal equipment to sell in the black market. Using GPS data, the managers wanted to build a machine learning model to identify suspicious driving activity. It's not a coincidence that minor tweaks to the same model would also allow management to monitor drivers' productivity: tracking how frequently they took bathroom breaks, for example, or whether they were sticking to the fastest possible routes.

The TCO managers were also interested in Microsoft products that could analyze large quantities of text. "Let's say we have the ability to mine everyone's emails," one executive asked. "What information could we find?"

When I reflect back on this meeting, it was a surreal experience. Everyone present discussed the idea of building a workplace panopticon with complete normalcy. The TCO managers claimed that monitoring workers was necessary for keeping them safe, or to prevent them from stealing. But it wasn't convincing in the slightest. We knew that they simply wanted a way to discipline their low-wage Kazakhstani workforce. We knew they wanted a way to squeeze as much work as they could from each worker.

I held my tongue and made sure to appear calm and collected. So did my colleagues. Collectively representing Microsoft, we turned a blind eye, and played along perfectly. We sympathized with TCO's incriminating portrayal of their Kazakhstani workers and the need to uphold the rule of law. We accepted their explanation that increased surveillance would improve worker safety. But truth be told, we didn't even need the excuses. Microsoft was hungry for their business. We were ready to concede.

Skip the Straw

The topic of worker surveillance took me by surprise. I didn't sign up for it. I did sign up for helping to accelerate the climate crisis, however — and it was something I had thought about a lot by the time I landed in Atyrau.

When I was first asked to present at the workshop, I was excited. It was good for my career, the technology was fascinating, and I had never been to Kazakhstan. But I hesitated. Did I really want to help Chevron destroy the planet? There were others on my team who could have easily gone in my place. Still, I decided to go. I wanted to learn about the oil industry and the kinds of investments that Big Tech was making. I wanted a front-row seat to the Microsoft-Chevron partnership. I wanted to know what we were up against.

During the workshop, I asked a coworker how she felt about Microsoft working with Big Oil. She responded sympathetically, understanding my concerns about climate change. But she also seemed to feel there was nothing we could do. For her and many other colleagues I've spoken to, change has to happen at the top. The problem, of course, is that the top has powerful incentives not to change. Microsoft executives aren't going to give up on the billions of dollars to be made from Big Oil, especially if it helps them win more of the coveted cloud market.

They are happy to offer employees small ways to live more sustainable lives, however. The company runs various recycling programs, encourages employees to "skip the straw" to reduce plastic consumption, and funds sustainability hackathons. (One hackathon project involved using AI/ML to detect trash in the ocean.) More broadly, Microsoft works hard to present an environmentally friendly public face. Its most ambitious green initiative is its promise to power its energy-hungry data centers with renewable sources. In 2016, Microsoft announced its goal to transition its data centers to 50 percent renewable energy by 2018. Hitting that target one year early, president and chief legal officer Brad Smith announced that the next goal is to surpass 70 percent renewable by 2023. "Time is too short, resources too thin and the impact too large to wait for all the answers to act," he said.

On the surface, then, Microsoft appears to be committed to fighting climate change. Google has constructed a similar reputation. But in reality, these companies are doing just enough to keep their critics distracted while teaming up with the industry that is at the root of the climate crisis. Why go through the effort of using clean energy to power your data centers when those same data centers are being used by companies like Chevron to produce more oil?

After Empire

At the workshop in Atyrau, a young Kazakhstani data scientist approached me to ask about a project that he was migrating to Microsoft's cloud platform. He didn't speak English fluently, but I could tell he was a good engineer. I wasn't sure if he really needed my help. It seemed like he just wanted to chat with another engineer in a room filled with businesspeople.

Afterwards, he told me a bit about how he ended up working for TCO, and how he wasn't able to find any other opportunities in the country that could match the offer. He had attended Purdue University to get an undergraduate degree in computer science. But since the Kazakhstan government paid for his tuition, he had to return to the country to work. "It means that I have to work in oil," he said. "It's basically the only industry that pays."

Speaking with him made me realize the extent of oil's dominance in Kazakhstan. Oil is by far the biggest economic sector, accounting for 63 percent of the country's total exports. In 2013, TCO made $15 billion in direct payments to the government—an enormous figure, considering that the country's entire tax revenue that year came to $21 billion. TCO is also a major source of wealth for the region. For years, the venture has invested millions of dollars into building schools, community centers, and fitness centers for the local people.

Kazakhstan's dependence on oil has only grown over the past decade. In 2016, TCO announced a $36.8 billion expansion to the Tengiz project, tying the country's economic future even more closely to fossil fuels. To make matters worse, the country's ability to produce oil relies heavily on multinational oil companies. At the time of its founding, TCO was a fifty-fifty partnership between Chevron and the state-owned KazMunayGas. Since then, ExxonMobil and the Russian oil company LukArco have joined the venture, but only KazMunayGas's share has been diluted.

While the country would struggle to take advantage of its oil-rich lands without the help of these foreign partners, the partnership is far from a win-win deal. Chevron keeps a tight grip on power, appointing most members of TCO's upper ranks. The power dynamic was clear at the workshop: lower-level employees were Kazahkstanis while management was almost entirely American. The local economy has also completely aligned itself with the needs of the American-dominated TCO. TCO proudly announced in Q1 of 2019 that it spent over $1 billion on Kazakhstani goods and services, which includes hiring more than forty thousand local workers to work in the oil field. But this makes local businesses highly dependent on TCO. If American oil companies pulled out of the venture or slashed funding, TCO would crumble, and many businesses would lose their biggest (and often only) customer, leaving the economy in shambles.

Big Tech isn't responsible for Kazakhstan's reliance on oil. Nor can we blame it for the climate catastrophe that we're facing. But it is certainly exacerbating both. While Kazahkstan's economy may benefit in the short run, intensifying the climate disaster will ultimately hurt the country too. Research shows that the region will suffer from increased aridity and more frequent heat waves, which could decrease crop yields and challenge food security.

How can tech help, instead of hurt, the climate? How can tech companies make local economies more resilient rather than more vulnerable? How can we demand climate justice from Microsoft, a company that claims to be a leader in the fight against climate change?

> *"If Chevron and other oil giants cease operations, it would decimate the economy of places like Kazakhstan — places whose dependency on oil has been actively encouraged by those companies, which have in turn profited handsomely from it."*

While I was in Atyrau, these very questions were being asked back home. Amazon employees in the US published an open letter calling on their company to reduce its carbon footprint and cancel its many contracts with Big Oil. Sitting in my hotel room not far from one of the largest oil fields in the world, I watched the letter blow up on my social media feeds. The number of Amazon signatures exploded: "3,500 employees challenge Bezos", "4,200 Amazon workers push for climate action", "6,000 employees sign an open letter to Bezos."

I was thrilled. Tech workers like me were taking a stand against our industry's role in accelerating the climate crisis. They weren't waiting for change at the top; they were demanding change from below.

Then I thought of the young Kazakhstani engineer. What happens to people like him after we decarbonize? If Chevron and other oil giants cease operations, it would decimate the economy of places like Kazakhstan — places whose dependency on oil has been actively encouraged by those companies, which have in turn profited handsomely from it. Resource extraction is an ancient imperial practice. As tech workers join the movement for climate justice, we must also find ways to undo the legacies of Big Oil's imperialism, and bring countries like Kazakhstan fairly and safely into a carbon-free future.

But it won't be easy. When I returned to the US, I learned that Bezos had effectively ignored the demands of over 8,000 of his employees. The open letter was an important first step, but more action will be needed for Amazon to drop its oil partnerships. We have a long fight ahead of us, and the stakes are high. We have, quite literally, a world to win. ∿

Zero Cool is the pseudonym of a software engineer at Microsoft.

The Body Instrumental

by Os Keyes

A technology that claims to recognize people's gender is becoming more widespread, with disastrous consequences.

———————

In 2019, Berlin celebrated Equal Pay Day by offering women discounts on public transit. It provided these discounts automatically, by analyzing the faces of people purchasing tickets. On the face of it, as it were, this approach might appear innocuous (or even beneficial — a small offset to gendered pay disparities!). But in actual fact, the technology in question is incredibly dangerous.

Automated Gender Recognition (AGR) isn't something most people have heard of, but it's remarkably common. A subsidiary technology to facial recognition, AGR attempts to infer the gender of the subject of a photo or video through machine learning. It's integrated into the facial recognition services sold by big tech companies like Amazon and IBM, and has been used for academic research, access control to gendered facilities, and targeted advertising. It's difficult to know all of the places where it's currently deployed, but it's a common feature of general facial recognition systems: anywhere you see facial recognition, AGR might well be present.

The growing pervasiveness of AGR is alarming because it has the potential to cause tremendous harm. When you integrate the assumptions embedded in this technology into our everyday infrastructure, you empower a system that has a very specific—and very exclusive—conception of what "gender" is. And this conception is profoundly damaging to trans and gender non-conforming people. AGR doesn't merely "measure" gender. It reshapes, disastrously, what gender means.

> *"There's only one small problem: inferring gender from facial features is complete bullshit."*

The Algorithmic Bathroom Bill

So what precisely is AGR, and where does it come from? The technology originated in academic research in the late 1980s (specifically in psychology—but that's another story) and started off with a particularly dystopian vision of the future it was creating. One early paper, after noting AGR's usefulness for classifying monkey faces, proposed that the same approach "could, at last, scientifically test the tenets of anthroposcopy (physiognomy), according to which personality traits can be divined from features of the face and head." Malpractice and harm have never been far from these systems.

When you run into a system that uses AGR, it takes a photograph (or video) of you, and then looks at your bone structure, skin texture, and facial shape. It looks at where (and how prominent) your cheekbones are, or your jawline, or your eyebrows. It doesn't need to notify you to do this: it's a camera. You may not even be aware of it. But, as it works out the precise points of similarity and difference between the features of your face and those of a template, it classifies your face as "male" or "female." This label is then fed to a system that logs your gender, tracks it, and uses it to inform the ads that an interactive billboard shows you or whether you can enter a particular gendered space (like a bathroom or a changing room).

There's only one small problem: inferring gender from facial features is complete bullshit.

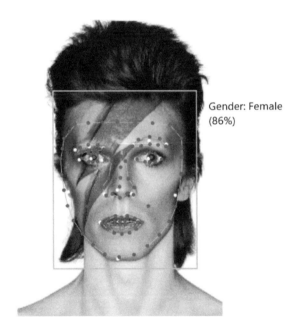

Gender: Female
(86%)

You can't actually tell someone's gender from their physical appearance. If you try, you'll just end up hurting trans and gender non-conforming people when we invariably don't stack up to your normative idea of what gender "looks like." Researchers such as myself and Morgan Scheuerman have critiqued the technology for precisely this reason, and Morgan's interviews with trans people about AGR reveal an entirely justified sense of foreboding about it. Whether you're using cheekbone structure or forehead shape, taking a physiological view of gender is going to produce unfair outcomes for trans people—particularly when (as is the case with every system I've encountered) your models only produce binary labels of "male" and "female."

The consequences are pretty obvious, given the deployment contexts. If you have a system that is biased against trans people and you integrate it into bathrooms and changing rooms, you've produced an algorithmic bathroom bill. If you have a situation that simply cannot *include* non-binary people, and you integrate it into bathrooms and changing rooms, you've produced an algorithmic bathroom bill.

A True Transsexual

So AGR clearly fails to measure gender. But why do I say that it *reshapes* gender?

Because all technology that implicates gender, alters it; more generally, all technology that measures a thing alters it simply by measuring it. And while we can't know all of the ramifications of a relatively new development like AGR yet, there are a ton of places where we can see the kind of thing I'm talking about. A prominent example can be found in the work of Harry Benjamin, an endocrinologist who was one of the pioneers of trans medicine.

Working in the 1950s, Benjamin was one of the first doctors to take trans people even marginally seriously, and at a vital time—the moment when, through public awareness of people like Christine Jorgensen (one of the first trans people to come out in the United States), wider society was first becoming seriously aware of trans people. While media figures argued back and forth about Jorgensen, Benjamin published *The Transsexual Phenomenon* in 1966, the first medical textbook about trans people ever written.

> "*Because all technology that implicates gender, alters it; more generally, all technology that measures a thing alters it simply by measuring it.*"

Containing case studies, life stories, diagnostic advice, and treatment approaches, *The Transsexual Phenomenon* became the standard medical work on trans subjects, establishing Benjamin as an authority on the matter. And it was, for its time, very advanced simply for treating trans medicine as a legitimate thing. It argued that trans people who wanted medical interventions would benefit from and deserved them, at a point when the default medical approach was "psychoanalyze them until they stop being trans." Benjamin believed this was futile, and that for those patients for whom it was appropriate, interventions should be made available.

To identify whether someone was such a patient, and what treatments should be made available, Benjamin built his book around his Sex Orientation Scale, often known simply as the Benjamin Scale.

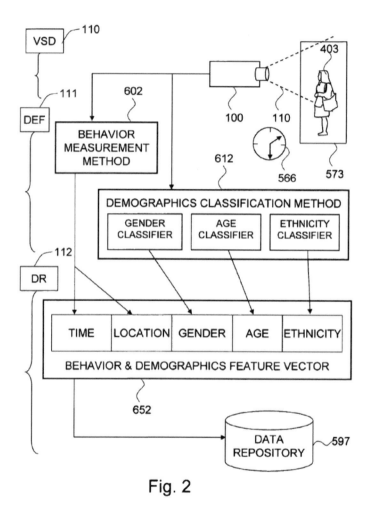

Fig. 2

"Automatic detection and aggregation of demographics and behavior of people," a patent for a system that includes AGR.

A doctor using the Benjamin Scale would first work to understand the patient, their life, and their state of mind, and then classify them into one of the six "types." Based on that classification, the doctor would determine what the appropriate treatment options might be. For Type V or Type VI patients, often grouped as "true transsexual," the answer was hormones, surgical procedures, and social role changes that would enable them to live a "normal life."

But this scale, as an instrument of measurement, came with particular assumptions baked into it about what it was measuring, and what a normal life was. A normal life was a heterosexual life: a normal woman, according to Benjamin, is attracted to men. A normal life meant two, and only two, genders and forms of embodiment. A normal life meant a husband (or wife) and a white picket fence, far away from any lingering trace of the trans person's assigned sex at birth, far away from any possibility of regret.

Further, it meant that trans women who were too "manly" in bone structure, or trans men too feminine, should be turned away at the door. It meant delay after delay after delay to ensure the patient *really* wanted surgery, advocating "a thorough study of each case... together with a prolonged period of observation, up to a year" to prevent the possibility of regret. It meant expecting patients to live as their desired gender for an extended period of time to ensure they would "pass" as "normal" after medical intervention—something known as the "real life test." Ultimately, Benjamin wrote, "the psychiatrist must have the last word."

So to Benjamin, a "true" trans person was heterosexual, deeply gender-stereotyped in their embodiment and desires, and willing to grit their teeth through a year (or more) of therapy to be sure they were *really* certain that they would prefer literally anything else to spending the rest of their life with gender dysphoria.

> *"When a technology assumes that men have short hair, we call it a bug. But when that technology becomes normalized, pretty soon we start to call long-haired men a bug."*

No More Ghosts

On its own, Benjamin's notion of a "normal life" would have been nothing but laughable — and laugh is what most of my friends do when I point them to the bit where he doesn't think queer trans people exist. But because of how widely his instrument of measurement has worked its way into systems of power, it has been deeply influential.

Benjamin's textbook and, more importantly, his scale, became standard in trans medicine, informing the design of the *Diagnostic and Statistical Manual of Mental Disorders* (DSM) definitions of gender dysphoria and the rules of the World Professional Association on Transgender Health (WPATH) — considered (by doctors) to be the gold standard in treatment approaches. Those rules still contain a "real life" test and psychiatric gatekeeping, and the DSM only began recognizing non-binary genders as real in 2013. More broadly, public narratives of transness still tell "the story" popularized and validated by Benjamin — the trans woman "born a girl, seeing herself in dresses," the trans man who has "always known" — even when that story does not and has never represented many of us.

The consequences for people who do not conform have been dire. People are denied access to medical care for not meeting the formal medical definition of a "true transsexual"; people are denied legitimacy in trans spaces for not "really" being trans; people are convinced by these discourses that their misery must be fake — that because they don't fit a particular normative idea of what a trans person is, they're not really a trans person at all, and so should go back into the closet for years or decades or the rest of their lives. All because of a tool that claimed merely to measure gender. Inside and outside our communities and selves, Benjamin's ghost continues to wreak unholy havoc.

So what is the point of this (admittedly fascinating) psychomedical history? The point is that there's no such thing as a tool of measurement that merely "measures." Any measurement system, once it becomes integrated into infrastructures of power, gatekeeping, and control, fundamentally changes the thing being measured. The system becomes both an opportunity (for those who succeed under it) and a source of harm (for those who fail). And these outcomes become naturalized: we begin to treat how the tool sees reality as reality itself.

When we look at AGR, we can observe this dynamic at work. AGR is a severely flawed instrument. But when we place it within the context of its current and proposed uses — when we place it within infrastructures — we begin to see how it not only measures gender but reshapes it. When a technology assumes that men have short hair, we call it a bug. But when that technology becomes normalized, pretty soon we start to call long-haired men a bug. And after awhile, whether strategically or genuinely, those men begin to believe it. Given that AGR developers are so normative that their research proposals include displaying "ads of cars when a male is detected, or dresses in the case of females," it's safe to say the technology won't be reshaping gender into something more flexible.

AGR might not be as flashy or obviously power-laden as the Benjamin Scale, but it has the potential to become more ubiquitous: responsive advertising and public bathrooms are in many more places than a psychiatrist's office. While the individual impact might be smaller, the cumulative impact of thousands of components of physical and technical reality misclassifying you, reclassifying you, punishing you when you fail to conform to rigid gender norms and rewarding you when you do, could be immense.

The good news is that the story of the Benjamin Scale shows us that resistance is possible. We did not go quietly into the psych; we fought, we lied, we hit back, and we continue to do so. But resistance is not enough. The norms that the Benjamin Scale worked into the world are still being perpetuated. Carving out space to breathe and live is good, but those battles are only necessary if you have already lost the war.

So rather than focus on reforming AGR—adding new categories or caveats or consent mechanisms, which are all moves that implicitly accept its deployment—we should push back more generally. We should focus on delegitimizing the technology *altogether*, ensuring it never gets integrated into society, and that facial recognition as a whole (with its many, many inherent problems) goes the same way. Do not just ask how we resist it—ask the people developing it why we *need* it. Demand that legislators ban it, organizations stop resourcing it, researchers stop designing it. Forty years after Benjamin's death, we are still haunted by his ghost. We don't need any more. 〜〜

Os Keyes is a researcher and essayist based at the University of Washington.

A Giant Bumptious Litter

Donna Haraway on Truth, Technology, and Resisting Extinction

The history of philosophy is also a story about real estate.

Driving into Santa Cruz to visit Donna Haraway, we can't help feeling that we were born too late. The metal sculpture of a donkey standing on Haraway's front porch, the dogs that scramble to her front door barking when we ring the bell, and the big black rooster strutting in the coop out back — the entire setting evokes an era of freedom and creativity that postwar wealth made possible in Northern California.

Here was a counterculture whose language and sensibility the tech industry sometimes adopts, but whose practitioners it has mostly priced out. Haraway, who came to the University of Santa Cruz in 1980 to take up the first tenured professorship in feminist theory in the US, still conveys the sense of a wide-open world.

Haraway was part of an influential cohort of feminist scholars who trained as scientists before turning to the philosophy of science in order to investigate how beliefs about gender shaped the production of knowledge about nature. Her most famous text remains "A Cyborg Manifesto," published in

1985. It began with an assignment on feminist strategy for the *Socialist Review* after the election of Ronald Reagan and grew into an oracular meditation on how cybernetics and digitization had changed what it meant to be male or female – or, really, any kind of person. It gained such a cult following that Hari Kunzru, profiling her for *Wired* years later, wrote: "To boho twentysomethings, her name has the kind of cachet usually reserved for techno acts or new phenethylamines."

The cyborg vision of gender as changing and changeable was radically new. Her map of how information technology linked people around the world into new chains of affiliation, exploitation, and solidarity feels prescient at a time when an Instagram influencer in Berlin can line the pockets of Silicon Valley executives by using a phone assembled in China that contains cobalt mined in Congo to access a platform moderated by Filipinas.

Haraway's other most influential text may be an essay that appeared a few years later, on what she called "situated knowledges." The idea, developed in conversation with feminist philosophers and activists such as Nancy Hartsock, concerns how truth is made. Concrete practices of particular people *make* truth, Haraway argued. The scientists in a laboratory don't simply observe or conduct experiments on a cell, for instance, but co-create what a cell is by seeing, measuring, naming, and manipulating it. Ideas like these have a long history in American pragmatism. But they became politically explosive during the so-called Science Wars of the 1990s – a series of public debates among "scientific realists" and "postmodernists" with echoes in controversies about bias and objectivity in academia today.

Haraway's more recent work has turned to human-animal relations and the climate crisis. She is a capacious *yes, and* thinker, the kind of leftist feminist who believes that the best thinking is done collectively. She is constantly citing other people, including graduate students, and giving credit to them. A recent documentary about her life and work by the Italian filmmaker Fabrizio Terranova, *Story Telling for Earthly Survival*, captures this sense of commitment, as well as her extraordinary intellectual agility and inventiveness.

At her home in Santa Cruz, we talked about her memories of the Science Wars and how they speak to our current "post-truth" moment, her views on contemporary climate activism and the Green New Deal, and why play is essential for politics.

Let's begin at the beginning. Can you tell us a little bit about your childhood?

I grew up in Denver, in the kind of white, middle-class neighborhood where people had gotten mortgages to build housing after the war. My father was a sportswriter. When I was eleven or twelve years old, I probably saw seventy baseball games a year. I learned to score as I learned to read.

My father never really wanted to do the editorials or the critical pieces exposing the industry's financial corruption or what have you. He wanted to write game stories and he had a wonderful way with language. He was in no way a scholar — in fact he was in no way an intellectual — but he loved to tell stories and write them. I think I was interested in that as well — in words and the sensuality of words.

The other giant area of childhood storytelling was Catholicism. I was way too pious a little girl, completely inside of the colors and the rituals and the stories of saints and the rest of it. I ate and drank a sensual Catholicism that I think was rare in my generation. Very not Protestant. It was quirky then; it's quirky now. And it shaped me.

How so?

One of the ways that it shaped me was through my love of biology as a materialist, sensual, fleshly being in the world as well as a knowledge-seeking apparatus. It shaped me in my sense that I saw biology simultaneously as a discourse and profoundly of the world. The Word and the flesh.

Many of my colleagues in the History of Consciousness department, which comes much later in the story, were deeply engaged with Roland Barthes and with that kind of semiotics. I was very unconvinced and alienated from those thinkers because they were so profoundly Protestant in their secularized versions. They were so profoundly committed to the disjunction between the signifier and signified — so committed to a doctrine of the sign that is *anti*-Catholic, not just non-Catholic. The secularized sacramentalism that just drips from my work is against the doctrine of the sign that I felt was the orthodoxy in History of Consciousness. So Catholicism offered an alternative structure of affect. It was both profoundly theoretical and really intimate.

Did you start studying biology as an undergraduate?

I got a scholarship that allowed me to go to Colorado College. It was a really good liberal arts school. I was there from 1962 to 1966 and I triple majored in philosophy and literature and zoology, which I regarded as branches of the same subject. They never cleanly separated. Then I got a Fulbright to go to Paris. Then I went to Yale to study cell, molecular, and developmental biology.

Did you get into politics at Yale? Or were you already political when you arrived?

The politics came before that—probably from my Colorado College days, which were influenced by the civil rights movement. But it was at Yale that several things converged. I arrived in the fall of 1967, and a lot was happening.

> "*I was way too pious a little girl, completely inside of the colors and the rituals and the stories of saints and the rest of it.*"

New Haven in those years was full of very active politics. There was the antiwar movement. There was anti-chemical and anti-biological warfare activism among both the faculty and the graduate students in the science departments. There was Science for the People [a left-wing science organization] and the arrival of that wave of the women's movement. My lover, Jaye Miller, who became my first husband, was gay, and gay liberation was just then emerging. There were ongoing anti-racist struggles: the Black Panther Party was very active in New Haven.

Jaye and I were part of a commune where one of the members and her lover were Black Panthers. Gayle was a welfare rights activist and the mother of a young child, and her lover was named Sylvester. We had gotten the house for the commune from the university at a very low rent because we were officially an "experiment in Christian living." It was a very interesting group of people! There was a five-year-old kid who lived in the commune, and he idolized Sylvester. He would clomp up the back stairs wearing these little combat boots yelling, "Power to the people! Power! Power!" It made our white downstairs neighbors nervous. They didn't much like us anyway. It was very funny.

Did this political climate influence your doctoral research at Yale?

I ended up writing on the ways that metaphors shape experimental practice in the laboratory. I was writing about the experience of the coming-into-being of organisms in the situated interactions of the laboratory. In a profound sense, such organisms are *made* but not *made up*. It's not a relativist position at all; it's a materialist position. It's about what I later learned to call "situated knowledges." It was in the doing of biology that this became more and more evident.

How did these ideas go over with your labmates and colleagues?

It was never a friendly way of talking for my biology colleagues, who always felt that this verged way too far in the direction of relativism.

It's not that the words I was using were hard. It's that the ideas were received with great suspicion. And I think that goes back to our discussion a few minutes ago about semiotics: I was trying to insist that the gapping of the signifier and the signified does not really determine what's going on.

But let's face it: I was never very good in the lab! My lab work was appalling. Everything I ever touched died or got infected. I did not have good hands, and I didn't have good passion. I was always more interested in the discourse, if you will.

But you found a supervisor who was open to that?

Yes, Evelyn Hutchinson. He was an ecologist and a man of letters and a man who had had a long history of making space for heterodox women. And I was only a tiny bit heterodox. Other women he had given space to were way more out there than me. Evelyn was also the one who got us our house for our "experiment in Christian living."

"He would clomp up the back stairs wearing these little combat boots yelling, 'Power to the people! Power! Power!' It made our white downstairs neighbors nervous."

God bless. What happened after Yale?

Jaye got a job at the University of Hawaii teaching world history and I went as this funny thing called a "faculty wife." I had an odd ontological status. I got a job there in the general science department. Jaye and I were also faculty advisers for something called New College, which was an experimental liberal-arts part of the university that lasted for several years.

It was a good experience. Jaye and I got a divorce in that period but never really quite separated because we couldn't figure out who got the camera and who got the sewing machine. That was the full extent of our property in those days. We were both part of a commune in Honolulu.

Then one night, Jaye's boss in the history department insisted that we go out drinking with him, at which point he attacked us both sexually and personally in a drunken, homophobic, and misogynist rant. And very shortly after that, Jaye was denied tenure. Both of us felt stunned and hurt. So I applied for a job in the History of Science department at Johns Hopkins, and Jaye applied for a job at the University of Texas in Houston.

"You write in a closed room while tearing your hair out of your head — it was individual in that sense. But then it clicks, and the words come, and you consolidate theoretical proposals that you bring to your community."

Baltimore and the Thickness of Worlding

How was Hopkins?

History of Science was not a field I knew anything about, and the people who hired me knew that perfectly well. Therefore they assigned me to teach the incoming graduate seminar: Introduction to the History of Science. It was a good way to learn it!

Hopkins was also where I met my current partner, Rusten. He was a graduate student in the History of Science department, where I was a baby assistant professor. (Today I would be fired and sued for sexual harassment — but that's a whole other conversation.)

Who were some of the other people who became important to you at Hopkins?

[The feminist philosopher] Nancy Hartsock and I shaped each other quite a bit in those years. We were part of the Marxist feminist scene in Baltimore. We played squash a lot — squash was a really intense part of our friendship. Her lover was a Marxist lover of Lenin; he gave lectures in town.

In the mid-to-late 1970s, Nancy and I started the women's studies program at Hopkins together. At the time, she was doing her article that became her book on feminist materialism, [*Money, Sex, and Power: Toward a Feminist Historical Materialism*]. It was very formative for me.

Those were also the years that Nancy and Sandra Harding and Patricia Hill Collins and Dorothy Smith were inventing feminist standpoint theory. I think all of us were already reaching toward those ideas, which we then consolidated as theoretical proposals to a larger community. The process was both individual and collective. We were putting these ideas together out of our struggles with our own work. You write in a closed room while tearing your hair out of your head — it was individual in that sense. But then it clicks, and the words come, and you consolidate theoretical proposals that you bring to your community. In that sense, it was a profoundly collective way of thinking with each other, and within the intensities of the social movements of the late 1960s and early 1970s.

The ideas that you and other feminist philosophers were developing challenged many dominant assumptions about what truth is, where it comes from, and how it functions. More recently, in the era of Trump, we are often told we are living in a time of "post-truth" — and some critics have blamed philosophers like yourselves for creating the

environment of "relativism" in which "post-truth" flourishes. How do you respond to that?

Our view was never that truth is just a question of which perspective you see it from. "Truth is perspectival" was never our position. We were against that. Feminist standpoint theory was always *anti*-perspectival. So was the Cyborg Manifesto, situated knowledges, [the philosopher] Bruno Latour's notions of actor-network theory, and so on.

> "The notion that you would or would not 'believe' in evolution already gives away the game."

"Post-truth" gives up on materialism. It gives up on what I've called semiotic materialism: the idea that materialism is always situated meaning-making and never simply representation. These are not questions of perspective. They are questions of worlding and all of the thickness of that. Discourse is not just ideas and language. Discourse is bodily. It's not embodied, as if it were stuck in a body. It's bodily and it's bodying, it's worlding. This is the opposite of post-truth. This is about getting a grip on how strong knowledge claims are not just possible but *necessary*—worth living and dying for.

When you, Latour, and others were criticized for "relativism," particularly during the so-called Science Wars of the 1990s, was that how you responded? And could your critics understand your response?

Bruno and I were at a conference together in Brazil once. Which reminds me: If people want to criticize us, it ought to be for the amount of jet fuel involved in making and spreading these ideas! Not for leading the way to post-truth. We're guilty on the carbon footprint issue, and Skyping doesn't help, because I know what the carbon footprint of the cloud is.

Anyhow. We were at this conference in Brazil. It was a bunch of primate field biologists, plus me and Bruno. And Stephen Glickman, a really cool biologist, a man we both love, who taught at UC Berkeley for years and studied hyenas, took us aside privately. He said, "Now, I don't want to embarrass you. But do you believe in reality?"

We were both kind of shocked by the question. First, we were shocked that it was a question of *belief*, which is a Protestant question. A confessional question. The idea that reality is a question of belief is a barely secularized legacy of the religious wars. In fact, reality is a matter of worlding and inhabiting. It is a matter of testing the holding-ness of things. Do things *hold* or not?

Take evolution. The notion that you would or would not "believe" in evolution already gives away the game. If you say, "Of course I believe in evolution," you have lost, because you have entered the semiotics of representationalism—and post-truth, frankly. You have entered an arena where these are all just matters of internal conviction and have nothing to do with the world. You have left the domain of worlding.

The Science Warriors who attacked us during the Science Wars were determined to paint us as social constructionists—that all truth is purely socially constructed. And I think we walked into that. We invited those misreadings in a range of ways. We could have been more careful about listening and engaging more slowly. It was all too easy to read us in the way the Science Warriors did. Then the right wing took the

Science Wars and ran with it, which eventually helped nourish the whole fake-news discourse.

Your opponents in the Science Wars championed "objectivity" over what they considered your "relativism." Were you trying to stake out a position between those two terms? Or did you reject the idea that either of those terms even had a stable meaning?

Both terms inhabit the same ontological and epistemological frame — a frame that my colleagues and I have tried to make hard to inhabit. Sandra Harding insisted on "strong objectivity," and my idiom was "situated knowledges." We have tried to deauthorize the kind of possessive individualism that sees the world as units plus relations. You take the units, you mix them up with relations, you come up with results. Units plus relations equal the world.

People like me say, "No thank you: it's relationality all the way down." You don't have units plus relations. You just have relations. You have worlding. The whole story is about *gerunds* — worlding, bodying, everything-ing. The layers are inherited from other layers, temporalities, scales of time and space, which don't nest neatly but have oddly configured geometries. Nothing starts from scratch. But the play — I think the concept of play is incredibly important in all of this — proposes something new, whether it's the play of a couple of dogs or the play of scientists in the field.

This is not about the opposition between objectivity and relativism. It's about the thickness of worlding. It's also about being of and for some worlds and not others; it's about materialist commitment in many senses.

To this day I know only one or two scientists who like talking this way. And there are good reasons why scientists remain very wary of this kind of language. I belong to the Defend Science movement and in most public circumstances I will speak softly about my own ontological and epistemological commitments. I will use representational language. I will defend less-than-strong objectivity because I think we have to, situationally.

Is that bad faith? Not exactly. It's related to [what the postcolonial theorist Gayatri Chakravorty Spivak has called] "strategic essentialism." There is a strategic use to speaking the same idiom as the people that you are sharing the room with. You craft a good-enough idiom so you can work on something together. I won't always insist on what I think might be a stronger apparatus. I go with what we can make happen in the room together. And then we go further tomorrow.

In the struggles around climate change, for example, you have to join with your allies to block the cynical, well-funded, exterminationist machine that is rampant on the earth. I think my colleagues and I are doing that. We have not shut up, or given up on the apparatus that we developed. But one can foreground and background what is most salient depending on the historical conjuncture.

Santa Cruz and Cyborgs

To return to your own biography, tell us a bit about how and why you left Hopkins for Santa Cruz.

Nancy Hartsock and I applied for a feminist theory job in the History of Consciousness department at UC Santa Cruz together. We wanted to share it. Everybody assumed we were lovers, which we weren't, ever. We were told by the search committee that they couldn't consider a joint application because they had just gotten this job okayed and it was the first tenured position in feminist theory in the country. They didn't want to do anything further to jeopardize it. Nancy ended up deciding that she wanted to stay in Baltimore anyway, so I applied solo and got the job. And I was fired from Hopkins and hired by Santa Cruz in the same week — and for exactly the same papers.

What were the papers?

The long one was called "Signs of Dominance." It was from a Marxist feminist perspective, and it was regarded as too political. Even though it appeared in a major journal, the person in charge of my personnel case at Hopkins told me to white it out from my CV.

The other one was a short piece on [the poet and novelist] Marge Piercy and [feminist theorist] Shulamith Firestone in *Women: a Journal of Liberation*. And I was told to white that out, too. Those two papers embarrassed my colleagues and they were quite explicit about it, which was kind of amazing. Fortunately, the people at History of Consciousness loved those same papers, and the set of commitments that went with them.

You arrived in Santa Cruz in 1980, and it was there that you wrote the Cyborg Manifesto. Tell us a bit about its origins.

It had a very particular birth. There was a journal called the *Socialist Review*, which had formerly been called *Socialist Revolution*. Jeff Escoffier, one of the editors, asked five of us to write no more than five pages each on Marxist feminism, and what future we anticipated for it.

This was just after the election of Ronald Reagan. The future we anticipated was a hard right turn. It was the definitive end of the 1960s. Around the same time, Jeff asked me if I would represent *Socialist Review* at a conference of New and Old Lefts in Cavtat in Yugoslavia [now Croatia]. I said yes, and I wrote a little paper on reproductive biotechnology. A bunch of us descended on Cavtat, and there were relatively few women. So we rather quickly found one another and formed alliances with the women staff who were doing all of the reproductive labor, taking care of us. We ended up setting aside our papers and pronouncing on various feminist topics. It was really fun and quite exciting.

Out of that experience, I came back to Santa Cruz and wrote the Cyborg Manifesto. It turned out not to be five pages, but a whole coming to terms with what had happened to me in those years from 1980 to the time it came out in 1985.

The manifesto ended up focusing a lot on cybernetics and networking technologies. Did this reflect the influence of nearby Silicon Valley? Were you close with people working in those fields?

It's part of the air you breathe here. But the real tech alliances in my life come from my partner Rusten and his friends and colleagues, because he worked as a freelance software designer. He did contract work for Hewlett Packard for years. He had a long history in that world: when he was only fourteen, he got a job programming on punch cards for companies in Seattle.

The Cyborg Manifesto was the first paper I ever wrote on a computer screen. We had an old HP-86. And I printed it on one of those daisy-wheel printers. One I could never get rid of, and nobody ever wanted. It ended up in some dump, God help us all.

The Cyborg Manifesto had such a tremendous impact, and continues to. What did you make of its reception?

People read it as they do. Sometimes I find it interesting. But sometimes I just want to jump into a foxhole and pull the cover over me.

In the manifesto, you distinguish yourself from two other socialist feminist positions. The first is the techno-optimist position that embraces aggressive technological interventions in order to modify human biology. This is often associated with Shulamith Firestone's book *The Dialectic of Sex* (1970), and in particular her proposal for "artificial wombs" that could reproduce humans outside of a woman's body.

Yes, although Firestone gets slotted into a quite narrow, blissed-out techno-bunny role, as if all her work was about reproduction without wombs. She is remembered for one technological proposal, but her critique of the historical materialist conditions of mothering and reproduction was very deep and broad.

> *"The established disorder of our present era is not necessary. It exists. But it's not necessary."*

You also make some criticisms of the ideas associated with Italian autonomist feminists and the Wages for Housework campaign. You suggest that they overextend the category of "labor."

Wages for Housework was very important. And I'm always in favor of working by addition not subtraction. I'm always in favor of enlarging the litter. Let's watch the attachments and detachments, the compositions and decompositions, as the litter proliferates. Labor is an important category with a strong history, and Wages for Housework enlarged it.

But in thinkers with Marxist roots, there's also a tendency to make the category of labor do too much work. A great deal of what goes on needs to be thickly described with categories *other* than labor — or in interesting kinds of entanglement with labor.

What other categories would you want to add?

Play is one. Labor is so tied to functionality, whereas play is a category of non-functionality.

Play captures a lot of what goes on in the world. There is a kind of raw opportunism in biology and chemistry, where things work stochastically to form emergent systematicities. It's not a matter of direct functionality. We need to develop practices for thinking about those forms of activity that are not caught by functionality, those which propose the possible-but-not-yet, or that which is not-yet but still open.

It seems to me that our politics these days require us to give each other the heart to do just that. To figure out how, with each other, we can open up possibilities for what can still be. And we can't do that in in a negative mood. We can't do that if we do nothing but critique. We need critique; we absolutely need it. But it's not going to open up the sense of what might yet be. It's not going to open up the sense of that which is not yet possible but profoundly needed.

The established disorder of our present era is not necessary. It exists. But it's not necessary.

Playing Against Double Death

What might some of those practices for opening up new possibilities look like?

Through playful engagement with each other, we get a hint about what can still be and learn how to make it stronger. We see that in all occupations. Historically, the Greenham Common women were fabulous at this. [*Eds.: The Greenham Common Women's Peace Camp was a series of protests against nuclear weapons at a Royal Air Force base in England, beginning in 1981.*] More recently, you saw it with the Dakota Access Pipeline occupation.

The degree to which people in these occupations *play* is a crucial part of how they generate a new political imagination, which in turn points to the kind of work that needs to be done. They open up the imagination of something that is not what [the ethnographer] Deborah Bird Rose calls "double death"—extermination, extraction, genocide.

Now, we are facing a world with all three of those things. We are facing the production of systemic homelessness. The way that flowers aren't blooming at the right time, and so insects can't feed their babies and can't travel because the timing is all screwed up, is a kind of forced homelessness. It's a kind of forced migration, in time and space.

This is also happening in the human world in spades. In regions like the Middle East and Central America, we are seeing forced displacement, some of which is climate migration. The drought in the Northern Triangle countries of Central America—Honduras, Guatemala, El Salvador—is driving people off their land.

So it's not a humanist question. It's a multi-kind and multi-species question.

In the Cyborg Manifesto, you use the ideas of "the homework economy" and the "integrated circuit" to explore the various ways that information technology was restructuring labor in the early 1980s to be more precarious, more global, and more feminized. Do climate change and the ecological catastrophes you're describing change how you think about those forces?

Yes and no. The theories that I developed in that period emerged from a particular historical conjuncture. If I were mapping the integrated circuit today, it would have different parameters than the map that I made in the early 1980s. And surely the questions of immigration, exterminism, and extractivism would have to be deeply engaged. The problem of rebuilding place-based lives would have to get more attention.

The Cyborg Manifesto was written within the context of the hard-right turn of the 1980s. But the hard-right turn was one thing; the hard-fascist turn of the late 2010s is another. It's not the same as Reagan. The presidents of Colombia, Hungary, Brazil, Egypt, India, the United States — we are looking at a new fascist capitalism, which requires reworking the ideas of the early 1980s for them to make sense.

So there are continuities between now and the map I made then, a lot of continuities. But there are also some pretty serious inflection points, particularly when it comes to developments in digital technologies that are playing into the new fascism.

Could you say more about those developments?

If the public-private dichotomy was old-fashioned in 1980, by 2019 I don't even know what to call it. We have to try to rebuild some sense of a public. But how can you rebuild a public in the face of nearly total surveillance? And this surveillance doesn't even have a single center. There is no eye in the sky.

Then we have the ongoing enclosure of the commons. Capitalism produces new forms of value and then encloses those forms of value — the digital is an especially good example of that. This involves the monetization of practically everything we do. And it's not like we are ignorant of this dynamic. We know what's going on. We just don't have a clue how to get a grip on it.

One attempt to update the ideas of the Cyborg Manifesto has come from the "xenofeminists" of the international collective Laboria Cuboniks. I believe some of them have described themselves as your "disobedient daughters."

Overstating things, that's not my feminism.

Why not?

I'm not very interested in those discussions, frankly. It's not what I'm doing. It's not what makes me vital now. In a moment of ecological urgency, I'm more engaged in questions of multispecies environmental and reproductive justice. Those questions certainly involve issues of digital and robotic and machine cultures, but they aren't at the center of my attention.

What is at the center of my attention are land and water sovereignty struggles, such as those over the Dakota Access Pipeline, over coal mining on the Black Mesa plateau, over extractionism everywhere. My attention is centered on the extermination and extinction crises happening at a worldwide level, on human and nonhuman displacement and homelessness. That's where my energies are. My feminism is in these other places and corridors.

> "*How can you rebuild a public in the face of nearly total surveillance? And this surveillance doesn't even have a single center. There is no eye in the sky.*"

Do you still think the cyborg is still a useful figure?

I think so. The cyborg has turned out to be rather deathless. Cyborgs keep reappearing in my life as well as other people's lives.

The cyborg remains a wily trickster figure. And, you know, they're also kind of old-fashioned. They're hardly up-to-the-minute. They're rather klutzy, a bit like R2-D2 or a pacemaker. Maybe the embodied digitality of us now is not especially well captured by the cyborg. So I'm not sure. But, yeah, I think cyborgs are still in the litter. I just think we need a giant bumptious litter whelped by a whole lot of really badass bitches — some of whom are men!

Mourning Without Despair

You mentioned that your current work is more focused on environmental issues. How are you thinking about the role of technology in mitigating or adapting to climate change — or

fighting extractivism and extermination?

There is no homogeneous socialist position on this question. I'm very pro-technology, but I belong to a crowd that is quite skeptical of the projects of what we might call the "techno-fix," in part because of their profound immersion in technocapitalism and their disengagement from communities of practice.

Those communities may need other kinds of technologies than those promised by the techno-fix: different kinds of mortgage instruments, say, or re-engineered water systems. I'm against the kind of techno-fixes that are abstracted from place and tied up with huge amounts of technocapital. This seems to include most geoengineering projects and imaginations.

So when I see massive solar fields and wind farms I feel conflicted, because on the one hand they may be better than fracking in Monterey County—but only maybe. Because I also know where the rare earth minerals required for renewable energy technologies come from and under what conditions. We still aren't doing the whole supply-chain analysis of our technologies. So I think we have a long way to go in socialist understanding of these matters.

One tendency within socialist thought believes that socialists can simply seize capitalist technology and put it to different purposes—that you take the forces of production, build new relations around them, and you're done. This approach is also associated with a Promethean, even utopian approach to technology. Socialist techno-utopianism has been around forever, but it has its own adherents today, such as those who advocate for "Fully Automated Luxury Communism." I wonder how you see that particular lineage of socialist thinking about technology.

I think very few people are that simplistic, actually. In various moments we might make proclamations that come down that way. But for most people, our socialisms, and the approaches with which socialists can ally, are richer and more varied.

When you talk to the Indigenous activists of the Black Mesa Water Coalition, for example, they have a complex sense around solar arrays and coal plants and water engineering and art practices and community movements. They have very rich articulated alliances and separations around all of this.

Socialists aren't the only ones who have been techno-utopian, of course. A far more prominent and more influential strand of techno-utopianism has come from the figures around the Bay Area counterculture associated with the *Whole Earth Catalog*, in particular Stewart Brand, who went on to play important intellectual and cultural roles in Silicon Valley.

They are not friends. They are not allies. I'm avoiding calling them enemies because I'm leaving open the possibility of their being able to learn or change, though I'm not optimistic. I think they occupy the position of the "god trick." [*Eds.: The "god trick" is an idea introduced by Haraway that refers to the traditional view of objectivity as a transcendent "gaze from nowhere."*] I think they are blissed out by their own privileged positions and have no idea what their own positionality in the world really is. And I think they cause a lot of harm, both ideologically and technically.

How so?

They get a lot of publicity. They take up a lot of the air in the room.

It's not that I think they're horrible people. There should be space for people pushing new technologies. But I don't see nearly enough attention given to what kinds of technological innovation are really needed to produce viable local and regional energy systems that don't depend on species-destroying solar farms and wind farms that require giant land grabs in the desert.

> *"There's not much mourning with the Stewart Brand types. There's not much felt loss of the already disappeared, the already dead."*

The kinds of conversations around technology that I think we need are those among folks who know how to write law and policy, folks who know how to do material science, folks who are interested in architecture and park design, and folks who are involved in land struggles and solidarity movements. I want to see us do much savvier scientific, technological, and political thinking with each other, and I want to see it get press. The Stewart Brand types are never going there.

Do you see clear limitations in their worldviews and their politics?

They remain remarkably humanist in their orientation, in their cognitive apparatus, and in their vision of the world. They also have an almost Peter Pan quality. They never quite grew up. They say, "If it's broken, fix it."

This comes from an incapacity to mourn and an incapacity to be finite. I mean that psychoanalytically: an incapacity to understand that there is no status quo ante, to understand that death and loss are *real*. Only within that understanding is it possible to open up to a kind of vitality that isn't double death, that isn't extermination, and which doesn't yearn for transcendence, yearn for the fix.

There's not much mourning with the Stewart Brand types. There's not much felt loss of the already disappeared, the already dead—the disappeared of Argentina, the disappeared of the caravans, the disappeared of the species that will not come back. You can try to do as much resurrection biology as you want to. But any of the biologists who are actually involved in the work are very clear that there is no resurrection.

You have also been critical of the Anthropocene, as a proposed new geological epoch defined by human influence on the earth. Do you see the idea of the Anthropocene as having similar limitations?

I think the Anthropocene framework has been a fertile container for quite a lot, actually. The Anthropocene has turned out to be a rather capacious territory for incorporating people in struggle. There are a lot of interesting collaborations with artists and scientists and activists going on.

The main thing that's too bad about the term is that it perpetuates the misunderstanding that what has happened is a human species act, as if human beings as a species necessarily exterminate every planet we dare to live on. As if we can't stop our productive and reproductive excesses.

Extractivism and exterminationism are not human species acts. They come from a situated historical conjuncture of about five hundred years in duration that begins with the invention of the plantation and the subsequent modeling of industrial capitalism. It is a situated historical conjuncture that has had devastating effects even while it has created astonishing wealth.

To define this as a human species act affects the way a lot of scientists think about the Anthropocene. My scientist colleagues and friends really do continue to think of it as something human beings can't stop doing, even while they understand my historical critique and agree with a lot of it.

It's a little bit like the relativism versus objectivity problem. The old languages have a deep grip. The situated historical way of thinking is not instinctual for Western science, whose offspring are numerous.

Are there alternatives that you think could work better than the Anthropocene?

There are plenty of other ways of thinking. Take climate change. Now, climate change is a necessary and essential category. But if you go to the circumpolar North as a Southern scientist wanting to collaborate with Indigenous people on climate change—on questions of changes in the sea ice, for example, or changes in the hunting and subsistence base—the limitations of that category will be profound. That's because it fails to engage with the Indigenous categories that are actually active on the ground.

There is an Inuktitut word, "sila." In an Anglophone lexicon, "sila" will be translated as "weather." But in fact, it's much more complicated. In the circumpolar North, climate change is a concept that collects a lot of stuff that the Southern scientist won't understand. So the Southern scientist who wants to collaborate on climate change finds it almost impossible to build a contact zone.

Anyway, there are plenty of other ways of thinking about shared contemporary problems. But they require building contact zones between cognitive apparatuses, out of which neither will leave the same as they were before. These are the kinds of encounters that need to be happening more.

A final question. Have you been following the revival of socialism, and socialist feminism, over the past few years?

Yes.

What do you make of it? I mean, socialist feminism is becoming so mainstream that even *Harper's Bazaar* is running essays on "emotional labor."

I'm really pleased! The old lady is happy. I like the resurgence of socialism. For all the horror of Trump, it has released us. A whole lot of things are now being seriously considered, including mass nonviolent social resistance. So I am not in a state of cynicism or despair. 〰

Seeing Carbon through Silicon

by Anne Pasek

Finding a future for the planet in the history of the microchip.

———

Climate action today is increasingly a question of exponents. Merely reducing greenhouse gases won't cut it; according to the UN Intergovernmental Panel on Climate Change, emissions must be halved in ten years and halved again in subsequent decades if we are to avoid the worst effects of global warming. Similarly, renewable energy needs to do more than just increase its market share; it must spread exponentially, replacing fossil fuel energy sources within fifty years.

These mandates represent an unprecedentedly rapid transition in the nature of our energy grids, transportation, housing, and all of the related patterns and habits that make up our daily lives. Decarbonization, if it is adequate to the climate math, must be both incredibly ambitious and incredibly disruptive.

Changes on this scale are difficult to imagine. To complicate matters, history offers scant examples for reference. Accordingly, the task of charting a pathway through decarbonization is in large part also a question of stretching our metaphorical imagination to reframe the possible. This is difficult, creative, and necessary work. It is also fraught with hazards.

In the global orbit of Silicon Valley thought, where disruption is a word with more positive cachet, one analogy is gaining momentum: we should think about carbon like we think about computers. The story of the microprocessor, after all, is a story of exponential growth curves and adoption rates: per Moore's Law, the density of semiconductors has doubled every two years, making computers cheaper, smaller, and more powerful — a win-win that fueled the digital revolution. Couldn't renewable energy follow a similar path?

> "Moore's Law is not a law of physics. It took considerable social effort and material happenstance to make the growth curve hold."

This idea is at the root of a sweeping policy proposal currently circulating in both UN climate conferences and Davos event halls: a global "Carbon Law," styled after Moore's Law, that sets a roadmap for exponential climate action. Unlike most views of climate change, this future is surprisingly optimistic. Carbon Law proponents point out that renewable energy, although currently representing only about 2 percent of global electricity generation, has already followed an exponential growth curve in its short history. They expect this trend to continue, given the proper incentives from governments and investment from industry. With the right social alignment, as they see it, the technology will simply take over.

The risk here, as with any framing comparison, is that the metaphor will not hold. Stories of digital disruption have long been sources of prediction, optimism, and analogy — as well as sites of dangerous fantasies. As a framework for energy transition, the Carbon Law can do more harm than good if it imparts the wrong lessons, provides false comfort, or seeks to mobilize the wrong people.

Metaphors matter: nothing less than the future of the planet is at stake. And interrogating the charisma of exponential thinking suggests that the Carbon Law is unlikely to help make that future a fair and habitable one. Insofar as silicon's history helps us understand carbon's future, its lessons are the opposite of those circulating at Davos. The story of silicon doesn't teach us to sit back, relax, and let technology save us. On the contrary: its real lesson is the power of purposeful struggle within systems of constraint.

Legislating Moore's Law

Exponential growth is remarkable wherever you find it, and the steady gains in chip densities that began in the late 1960s remains a defining standard for rapid technological advancement. Moore's Law, however, is not a law of physics. It took considerable social effort and material happenstance to make the growth curve hold.

In 1965, Gordon E. Moore was the director of research at Fairchild Semiconductor, a pioneering firm that helped create Silicon Valley. In a magazine article, he observed that the number of components in integrated circuits had grown exponentially over the past seven years and would likely continue on this trajectory a decade hence. This prediction could have easily faltered. At the time, shrinking transistor sizes seemed to many engineers to invite disaster through unneeded complexity and melted components. Miniaturization, moreover, was an imperative unique to military contracts that needed chips small enough to fit onto rockets, while researchers were quite content to have room-sized computers.

It took considerable barn-raising by key figures in the industry, as well as hefty military spending, to make Moore's prediction into a de facto law by changing industry R&D allocations and targets. The ensuing rates of growth were formally ratified in national and international industry roadmaps in the 1990s, essentially securing Moore's Law as a group-fulfilling prophecy.

From the start, a combination of peer organizing and institutional mandates propelled Moore's supposition into a standard. The technology did not simply take over.

Yet this is not to say that exponential doubling is a purely socially constructed outcome. The unique properties of silicon supported the otherwise unlikely win-win of densification and miniaturization. The price, size, and processing power of chips are tightly correlated; smaller silicon circuits require less energy to power, produce less heat, and can process at faster speeds.

Writ large as an industry-wide trend, this fact about silicon's thermo-electric properties led to the massive popularization of cheaper, smaller, and more powerful digital devices. But this isn't true of electrical grids, photovoltaics, or other forms of technology that don't experience consistent density scaling across their components. If, for example, Moore's Law applied to air travel, a New York City-to-Stockholm flight today would hold 120 million passengers and take eight minutes.

However, even within the scalar logics of silicon, the predictive success of Moore's Law today is widely acknowledged to be over. Microchips are heating up, R&D costs threaten to outpace density gains, and, as engineers parse the design challenges of nanometer circuits, they may simply be running out of atoms. Future prospects for densification are multiple and uncertain. Rather than relying on exponential growth in processing capacities, software designers are increasingly depending on gains in efficiency first developed for mobile applications — seeking as an industry, if somewhat belatedly, to do more with less.

> *"If Moore's Law applied to air travel, a New York City-to-Stockholm flight today would hold 120 million passengers and take eight minutes."*

Surfing the Waves of the World Spirit

Techno-optimism is easy when exponential growth holds. Proponents of the Carbon Law largely see technology — both digital and electric — in this register. As a result, exponential decarbonization appears to them as little more than a technological coordination problem. It requires innovation and cooperation on the part of politicians and green tech companies, but asks very little from citizens. Its model of power is predicated on the agency of executives and devices, not political mobilization by large numbers of people.

Two men sit at the center of the Carbon Law's development and distribution, and their backgrounds help explain some of its political character: Johan Rockström, director of the Stockholm Resilience Center, and Johan Falk, the director of

Intel's Stockholm IoT Ignition Lab until he quit to work with Rockström. At Intel, Falk's mandate was to promote the spread of smart networks to new industries—to spread narratives of exponential growth and scalar disruption. Rockström's center, on the other hand, develops and disseminates "resilience thinking" across high-profile climate talks and a C-suite executive education program. Together, they have targeted political and corporate elites with a simple message: the climate crisis can't be solved without exponential thinking, which requires elite "accelerators" to reenact the roadmap and feedback loops that propelled Moore's Law forward.

> "*Decarbonization will demand more than just a different kind of technology curve, accelerating sharply into the horizon.*"

Strip away the Silicon Valley language from this proposal and the details are themselves common enough to most mainstream climate governance plans: price carbon, make cuts across multiple sectors, increase energy efficiency, and fund renewable R&D. What's unusual here is the emotional register of the plan and the apolitical certainty of its promise—factors that can't be disentangled from tech industry tropes. For instance, Falk's "Exponential Roadmap Report" argues that decarbonization "is nothing short of a global economic transformation. But transformation appears assured through revolutions driven by digitalisation. Harvesting this power will help drive unstoppable momentum." Similarly, Rockström predicts that under the Carbon Law, "big masses

[will] simply surf along a sustainable journey without knowing that they're doing it."

In short, the technology will do the work. Exponential growth curves will continue along an unchanging trajectory, as if by natural law. Existing social arrangements, fossil fuel interests, and the economic and environmental justice barriers to the energy transition will cede to the power of elite leadership and digital disruption. In turn, our carbon footprints, seemingly without effort, will just shrink and shrink.

Creativity Within Constraints

This top-down, technologically determined future ignores all the ways in which energy transitions aren't just a question of market shares, but of the social pressures and material constraints that cut across them. Decarbonization will demand more than just a different kind of technology curve, accelerating sharply into the horizon. It will very likely require abrasive changes to well-worn cultural norms, the structure of cities and trade, and perhaps even the valorization of economic growth in its broadest terms. It will be conflictual, classed, and expensive.

Technology alone proves to be a poor analytic for these kinds of social changes. Moreover, as demonstrated by recent waves of popular opposition to climate policy, market fixes without considerations for equity are politically disastrous. People, infrastructure, and culture don't fit into industry roadmaps or silicon wafers. They contain differences and resistances that can't be universally scaled.

If Moore's Law is to be a useful story through which to approach this future, it will be for all the reasons its green proponents currently ignore. The history of the microprocessor revolution is ultimately about the immensity of effort that goes into maintaining the dream of exponential growth—and its inevitable collapse. Moore's Law was neither a socially constructed prophecy nor a materially determined outcome. It was a period of coordinated action within specific material parameters that have now passed. It leaves us facing a technological future that will require creativity within new constraints.

Rockström and Falk are correct that time is short and the need to muster political and technological resources is great. Where they are wrong is the assumption that a better future will arrive on our desks without a fight. Instead, it will require a public that can stand up and push. ∿

Anne Pasek is a postdoctoral fellow in Transitions in Energy, Culture, and Society at the University of Alberta.

From Manchester to Barcelona

by Ben Tarnoff

Building a better story about the internet.

————

For a long time, a certain set of assumptions dominated our digital imagination. These assumptions should be familiar enough. Information wants to be free. Anything that connects people is good. The government is bad. The internet is another world, where the old rules don't apply. The internet is a place of individual freedom, which is above all the freedom to express oneself.

Such ideas were never 100 percent hegemonic, of course. They were always contested, with varying degrees of success. Governments, for one, found several ways to assert their sovereignty over online spaces. Scholars sounded the alarm on the rise of the white supremacist web—the notorious neo-Nazi site Stormfront launched in 1996—and presciently observed that the internet's connectivity could also make the world worse.

Even so, these assumptions and the intellectual traditions they emanated from—techno-utopianism, cyberlibertarianism, the

Californian Ideology—largely kept their grip on the common sense. The long 1990s is said to have begun with the fall of the Berlin Wall in 1989 and ended with the attacks of September 11, 2001. But, when it came to our popular discourse about the internet, the long 1990s lasted a lot longer.

Then came Snowden. In 2013, the former NSA contractor revealed that the internet was a vast spy machine for the American security state. A tremor of tech pessimism crept into public consciousness. Then came Trump. The media's failure to anticipate the possibility of his victory in 2016 led it to amplify the significance of Russian influence operations via social media—operations that clearly existed, but which, at a moment of supreme disorientation, metastasized into the deus ex machina that could explain an inexplicable result. Yet this coping mechanism had a silver lining: it provided the initial spark for what has come to be known as the "techlash."

"The long 1990s are over. The old gods are finally dead."

Journalists and politicians began to pay closer, less credulous attention to the internet and the companies that control it. Disinformation remained a key concern, but far from the only one: a long series of tech scandals have fed the fire, too many to keep count. The right has also joined the fray: the (laughable) notion that the big platforms silence conservative voices has taken root in the reactionary mind, turning a range of right-wing figures into harsh critics of Silicon Valley.

The resulting shift is stark. A sharper tone prevails in the *New York Times* and on Fox News, in statehouses and on Capitol Hill.

Criticisms once confined to scholarly circles, or to more oppositional outlets like *The Baffler* and Valleywag, have become conventional, even banal. One could be uncharitable about the heavy Kool-Aid drinkers who abruptly sobered up—there is no shortage of annoying figures among the late converts to tech critique—but the techlash has been a very good thing. We are at last having a more honest conversation about the internet. The long 1990s are over. The old gods are finally dead.

Who are the new gods? This is what makes our moment so interesting: the conventional wisdom is cracking up but its replacement hasn't quite consolidated. As James Bridle says, something is wrong on the internet—and something is wrong with the way we have thought about the internet—but there is not yet a widely accepted set of answers to the all-important questions of why these things are wrong, or how to make them right.

Different camps are now competing to provide those answers. They are competing to tell a new story about the internet, one that can explain the origins of our present crisis and offer a roadmap for moving past it. Some talk about monopoly and antitrust. Others emphasize privacy and consent. Shoshana Zuboff proposes the term "surveillance capitalism" to describe the new kinds of for-profit monitoring and manipulation that the internet and associated technologies have made possible.

These analyses have important differences. But they tend to share a liberal understanding of capitalism as a basically beneficent system, if one that occasionally needs state intervention to mitigate its excesses. They also tend to equate capitalism with markets. Sometimes these markets become too consolidated and need to be made more competitive (the antitrust view); sometimes market actors violate the terms of fair exchange and need to be restrained (Zuboff's view). But two articles of faith always remain. The first is that capitalism is more or less compatible with people's desire for dignity and self-determination (or can be made so with proper regulation). The second is that capitalism is more or less the same thing as markets.

What if neither belief is true? This is the starting point for building a better story about the internet.

> *"What makes capitalism so unusual is that production (and accumulation) isn't for anything exactly, aside from making it possible to produce (and accumulate) more."*

The Archipelago and the Network

If capitalism isn't (only) markets, then what is it?

There have always been markets. Capitalism, by contrast, is relatively new. Its laws of motion first emerged in Europe in the fifteenth and sixteenth centuries, and reached escape velocity with industrialization in the eighteenth and nineteenth.

If capitalism didn't invent markets, however, it did make markets much more important. The historian Robert Brenner observes that capitalism is defined above all by market dependence. Pre-capitalist peasants can trade and barter, but they don't depend on the market for life's necessities: they grow their own food. In capitalist societies, on the other hand, the market mediates your access to the means of subsistence. You must buy what you need to survive, and to have the money to do so, you must sell your labor power for a wage.

Market dependence doesn't exist for its own sake. It serves an important function: to facilitate accumulation. Accumulation is the aim of any capitalist arrangement: to take a sum of value and make more value out of it. While markets are certainly central to capitalism, they aren't what makes it tick. Accumulation is. To put it in a more Marxist idiom, capital is value in motion. As it moves, it expands. Capitalism, then, is a way to organize human societies for the purpose of making capital move.

There are a few different methods for making capital move. The principal one is for capitalists to purchase people's labor power, use it to create new value in the form of commodities, and then realize that value as profit by selling those commodities. A portion of the proceeds are reinvested into expanding production, so even more commodities can be made at lower cost, thus enabling our capitalist to compete effectively with the other capitalists selling the same commodities.

This may seem entirely obvious, but it's actually a very distinctive way of doing things. In other modes of social organization, the point of production is to directly fulfill people's needs: think of subsistence farmers, growing food for their families to eat. Or the point is to make the rulers rich: think of the slaves of ancient Rome, doing the dirty work so that imperial elites could lead lives of luxury.

What makes capitalism so unusual is that production (and accumulation) isn't for anything exactly, aside from making it possible to produce (and accumulate) more. This obsession gives capitalism its extraordinary dynamism, and its revolutionary force. It utterly transforms how humans live and, above all, how they produce. Capitalism forces people to produce together, in increasingly complex combinations of labor. Production is no longer solitary, but social.

This dynamic is most vividly illustrated by the factory. The modern factory was largely born in nineteenth-century Manchester, where Friedrich Engels's father co-owned a cotton mill. This gave the young Engels the opportunity to observe the birth of the factory up close. He saw hundreds, even thousands of workers, crammed into vast buildings, arrayed around machines, and performing different roles within a complex division of labor in order to work as one. What they made, they made together.

In pre-capitalist Europe, one person or a few people could plausibly claim credit for producing something. This wasn't the case in the capitalist factory, however. "The yarn, the cloth, the metal articles that now come out of the factory were the joint product of many workers, through whose hands they had successively to pass before they were ready," Engels wrote. "No one person could say of them: 'I made that; this is my product.'"

Yet there was a contradiction lurking here. If no one worker could claim sole credit for a product, the owner of the factory could still claim sole ownership of everything the workers made together. Wealth was being created socially, on a new model—but still owned privately, on the old model.

The contradiction became even sharper when zooming out to consider the wider economy. As many workers as it took to run a Manchester mill, it took even more workers to make that work possible, from the machinists who manufactured and maintained the power looms and the other machines to the slaves in the American South who picked the cotton that kept those machines fed. The collective labor inside the mill was sustained by many concentric circles of collective labor outside of it.

The pre-capitalist economy looked like a cluster of islands — an archipelago. It involved a collection of small producers relatively isolated from one another and producing mostly for personal use. (Marx memorably compared the French peasantry to a sack of potatoes.) By contrast, the capitalist economy looked like a network. The network of capital concentrated masses of people into larger nodes of production and linked them through countless threads of interdependence. Yet the wealth that this network generated didn't flow to the many workers who collectively created that wealth. It flowed to the few who owned the network: the capitalists.

Before capitalism, when production happened on a more personal basis, such an arrangement might've made sense. If the economy was a cluster of islands, it followed that each island would own what it made. But capitalism, by revolutionizing production, introduced a contradiction: wealth was now *made* as a network, but still *owned* as an archipelago. Capitalists like Engels's father became rich. The workers of Manchester earned starvation wages, and lived in cholera-infested slums.

The New Manchester

What does this have to do with the internet?

The internet, and the constellation of digital technologies that we call "tech" more broadly, intensifies the fundamental contradiction in capitalism between wealth being collectively produced and privately owned. It takes the Manchester model and elevates it to the nth degree. It makes the creation of wealth more collective than ever before, piling up vast new fortunes in the process — fortunes that, as they did in Engels's day, accrue to a small handful of owners.

A worker in a Manchester mill couldn't point to a finished piece of yarn and say, "I made that," but a few thousand workers (and slaves) probably could. Tech's wealth, on the other hand, is woven out of the contributions of *billions* of people, living and dead.

This helps explain why the tech industry is so ludicrously profitable. Take Facebook. In 2018, Facebook reported a net income of $22 billion with an operating margin of 45 percent. The company only has about 40,000 full-time employees, along with an undisclosed number of contractors. In other words, relative to its costs, Facebook makes an absurd amount of money. And Facebook's power isn't just about money: as the dominant media ecosystem in many countries around the world, it also embodies what Frank Pasquale calls "functional sovereignty." It operates like a government—which is particularly evident in the case of Libra, its new global digital currency. And this government is quite explicitly autocratic given a shareholder structure that preserves Mark Zuckerberg's personal control of the company.

It's hard to imagine a more extreme form of the contradiction on display at Manchester than a social network of more than two billion people ruled by a single billionaire. The network of capital has become denser, and more literal, than Engels could've possibly imagined, while its control has become concentrated in even fewer hands.

To observe that Facebook has relatively few workers is not to suggest that the work they perform is not important. Without content moderators, data center technicians, site reliability engineers, and others, Facebook's product would become unusable and its business would collapse. But their collective labor, like that of the workers within Engels's father's factory, depends on many concentric circles of collective labor outside of it. And, for Facebook and the other firms that fall under the umbrella of tech, the share of value supplied by these external layers is especially vast.

One source is the workers who invented the software, hardware, protocols, and programming languages that laid the basis for today's tech industry. These were developed over the course of several decades, starting with the creation of the first modern electronic computers in the 1940s, and relied heavily, often exclusively, on US military funding. Another source is the workers who, in the present day, continue to make and maintain the stuff on which tech profits depend. While this work takes many forms, most of it is menial or dangerous. It includes manufacturing smartphones, mining rare earth elements, and labeling training data for machine learning models.

"*Tech's wealth is woven out of the contributions of* billions *of people, living and dead.*"

As varied as these jobs are, though, they still look like traditional labor. People work and get paid, whether they're inventing the internet protocols or laying fiber-optic cable. Tech, however, also manages to draw value from activities that don't look like traditional labor. To return to Facebook, those more than two billion users create value for the company by supplying the site with its posts, comments, and likes. This content, paired with the rest of their activity on the platform, also furnishes Facebook with the personal data it uses to sell targeted advertising, which makes up the vast majority of its revenue.

It's a contested theoretical question whether all this posting and clicking should count as "labor" — and if so, what kind. In her canonical article on the subject, the theorist Tiziana Terranova uses the term "free labor" to describe the various unwaged activities that propped up profit-making in the early days of the commercial internet, from volunteer moderators on America Online to open-source software developers. But the scope of such activities has grown dramatically since Terranova published her piece in 2000, and they look less and less like labor. Increasingly, tech is able to harvest value from us simply for existing.

"The new Manchester is everywhere."

A good example comes in the form of a cafe in San Francisco called Brainwash. This cafe, since closed, had a camera inside of it that filmed customers. A group of researchers obtained the footage, and turned it into a dataset to train machine learning models for detecting heads and faces. Published in 2016, this dataset was then used by the Chinese firm Megvii, a global leader in facial recognition, to refine its own software. Megvii also happens to be implicated in the totalitarian surveillance state that the Chinese Communist Party is constructing in the western province of Xinjiang. In other words, by walking into a cafe one day in San Francisco, you may have helped a tech company make money by selling the Chinese government a product it uses to repress millions of its citizens some six thousand miles away. (Megvii is currently valued at $4 billion, and hopes to raise as much as $1 billion in an IPO expected for late 2019.)

These kinds of strange and tangled value chains will only become more common in the coming years. As small networked computers burrow ever more deeply into our homes, stores, streets, and workplaces, more data will be made. Meanwhile, advances in machine learning and the growth of cloud-based processing power will continue to make data more valuable, as the fuel that feeds automated systems for everything from recognizing faces to predicting consumer preferences.

The upshot is a world where the creation of wealth becomes more collective than ever before. In the nineteenth century, Engels reflected on how capitalism transformed production "from a series of individual into a series of social acts." The total enclosure of our world by computing means that those social acts can now happen at the scale of entire societies. The industrial factory has become what Terranova and others, building on a term from Italian autonomism, call the "social factory." The new Manchester is everywhere.

The Difference Engine

Capitalism connects. In its perpetual push to accumulate, it draws people into new sites and circuits of collective wealth-making. But if capitalism is a connector, it is also a differentiator. If capitalism is a network for making wealth, it is also an engine for making difference.

To watch this differentiating dynamic at work, let's return to Manchester for a moment. The people who collectively created the city's wealth were not a single homogenous mass. Quite the opposite: they were divided into men and women, English and Irish, white and Black. And these divisions were constantly being reinforced, since they served a valuable purpose: they helped make exploitation seem justified, even natural.

Thus it was natural for the Irish to be paid less and live in appalling slums. It was natural for women to be paid less while also performing the unpaid work of raising children — children who went into the mills as young as five. It was natural to enslave human beings of African origin and put them to work harvesting the cotton that those mills turned into textiles. It was natural to dispossess and exterminate the Indigenous people who had formerly inhabited the land that became those cotton fields.

"*The network for making wealth, in other words, relies on the engine for making difference.*"

Capitalism doesn't invent human difference, of course. Humans look different; they speak different languages; they come from different communities and cultures. But capitalism makes these differences make *more* of a difference to people's lives. Differences become more differential. They become differences of capacity and value — differences in how much a human being is worth, or if they're even considered human at all.

The political scientist Cedric J. Robinson argued that this difference-making has been a core feature of capitalism from the beginning — he called it *racial* capitalism for this reason. Feudal Europe was highly racialized, Robinson said. As Europeans conquered and colonized one another, they came up with ideas about racial difference in order to justify why, for instance, Slavs should be slaves. (In fact, Slavs were so frequently enslaved in the Middle Ages that they supplied the source of the word "slave," in English and several other European languages.)

If racial thinking saturated the societies where capitalism first emerged, capitalism subsequently picked up these concepts and extended them. It generated deeper and more varied ideas about racial difference in order to justify the new relationships of domination that the imperative of accumulation demanded — particularly as Europeans began carving up Asia, Africa, and the Americas. "The tendency of European civilization through capitalism," Robinson wrote, "was thus not to homogenize but to differentiate — to exaggerate regional, subcultural, and dialectical differences into 'racial' ones."

Robinson's insight helps clarify another crucial aspect of how tech operates. If tech intensifies capitalism's contradiction between wealth being collectively produced and privately owned, it also intensifies capitalism's tendency to slice people into different groups and assign them different capacities and values. Indeed, the two operations are closely related. "Capital can only be capital when it is accumulating," says the theorist Jodi Melamed, "and it can only accumulate by producing and moving through relations of severe inequality among human groups." The network for making wealth, in other words, relies on the engine for making difference.

That engine is now made of software. Differentiation happens at an algorithmic level. The abundant data that flows from mass digitization, combined with the ability of machine learning algorithms to find patterns in that data, has given capitalism vastly more powerful tools for segmenting and sorting humanity.

Way back in 1993, the media scholar Oscar H. Gandy, Jr. offered an extremely prescient view of how this works. He called it "the panoptic sort," in a book of the same name. "The panoptic sort is a difference machine that sorts individuals into categories and classes on the basis of routine measurements," he wrote. "It is a discriminatory technology that allocates options and opportunities on the basis of those measures and the administrative models that they inform."

Gandy was looking at how corporations and governments collected and processed personal information at a time when computing was widespread, but fairly primitive by today's standards—the commercial internet was still years away. Even so, Gandy discerned a logic that by now feels very familiar. Data was being drawn from many sources—thus the "panoptic" part—in order to sort people "according to their presumed economic or political value." And this operation wasn't peripheral or incidental to capitalism, but absolutely integral to it: the panoptic sort, Gandy argued, was "the all-seeing eye of the difference machine that guides the global capitalist system."

Today, this all-seeing eye sees much, much more. And the stakes of the sorting are even higher. Algorithmic differentiation helps determine who gets a loan, who gets a job, who goes to jail. Moreover, Gandy observed how the panoptic sort amplified existing disparities, racial and otherwise. This is far truer today, thanks to the mainstreaming of machine learning systems.

In recent years, scholars and journalists have drawn attention to the problem of "algorithmic bias." Such bias is endemic to machine learning because it "learns" by training on data drawn from our social world—data that inevitably reflects centuries of capitalist difference-making. Thus "predictive policing" algorithms trained on data that shows that the police arrest a lot of Black people suggest arresting more Black people. Or an Amazon algorithm trained on the resumes of its mostly male workforce advises against hiring women.

The role of these systems is not just to reproduce inequalities, but to naturalize them. Capitalist difference-making has always required a substantial amount of ideological labor to sustain it. For hundreds of years, philosophers and priests and scientists and statesmen have had to keep saying, over and over, that some people really are less human than others—that some people deserve to have their land taken, or their freedom, or their bodies ruled over or used up, or their lives or labor devalued. These ideas do not sprout and spread spontaneously. They must be very deliberately transmitted over time and space, across generations and continents. They must be taught in schools and churches, embodied in laws and practices, enforced in the home and on the street.

It takes a lot of work. Machine learning systems help automate that work. They leverage the supposed authority and neutrality of computers to make the differences generated by capitalism look like differences generated by nature. Because a computer is saying that Black people commit more crime or that women can't be software engineers, it must be true. To paraphrase one right-wing commentator, algorithms are just math, and math can't be racist. Thus machine learning comes to automate not only the production of inequality but its rationalization.

The New Barcelona

Anything that moves has an ideal medium for its motion. A fish moves best in water; a car moves best on pavement. Capital is value in motion, so it must always be moving. And it moves best through a particular kind of social fabric, one that is both webbed and fissured, linked and sliced, connected and differentiated.

This helps make sense of what we call tech. Tech is an agent and accelerant of these dynamics, of "densely connected social separateness," to borrow a term from Melamed. This explains its tendency to generate immense imbalances of wealth and power, and to heighten the hierarchical sorting of human beings according to race, gender, and other categories.

For our analysis to be useful, though, it needs to have not only a descriptive but a prescriptive element. It needs to offer some answers to the question of what is to be done.

This is where things get murkier, as one might expect. But there is clarity on at least one point. If tech intensifies the contradiction between wealth being made by the many and owned by the few, then the obvious solution is to resolve the contradiction: to turn socially made wealth into socially *owned* wealth. Or, as Marx and Engels put it in *The Communist Manifesto*, to convert the "collective product" of capital into "common property, into the property of all members of society."

The logic is appealingly simple: if the network makes the wealth, then let the network own the wealth. But how, precisely? What does it mean to transform the wealth that society makes in common into the common property of society? This is the most bitterly debated question in the whole history of the radical left. For most of the actually existing socialisms of the twentieth century, the answer was full nationalization on the Soviet model. This answer hasn't aged well.

Another approach, and one that is currently enjoying renewed popularity, draws from the tradition of worker self-management. This tradition comes in many flavors, but perhaps its most heroic moment occurred in revolutionary Catalonia during the Spanish Civil War, when people seized factories, farms, even flower shops and, for a brief period, ran everything themselves.

A young Marxist from Kentucky named Lois Orr would later remember the thrill of strolling through anarchist Barcelona and seeing its "cafés, restaurants, hotels, and theaters lit up red or red and black [with] banners saying Confiscated, Collectivized."

Barcelona, then, is one alternative to Manchester. But what would self-management mean for tech? A number of different experiments offer preliminary materials towards an answer. There are small, cooperatively owned platforms for everything from ride-hailing to social media. There are municipally owned broadband networks governed by local communities. There is an initiative to create a socially owned smart city in, of all places, Barcelona. There are also more ambitious but less immediately feasible schemes for democratizing the big platforms, whether by converting them into cooperatives of some kind or socializing their data.

These projects and proposals have the virtue of being concrete. As working hypotheses, they are immensely valuable. But they remain necessarily incomplete and provisional, particularly when considered as possible directions for moving beyond capitalism. Cooperatives under capitalism often behave like normal firms, since they are subject to the same market imperatives as everyone else. There is no straight line, then, from experiments in self-management to the broader goal of breaking with the logic of infinite accumulation and rebuilding society on a radically different basis.

Neither is there a direct relationship between democratizing ownership and combating the various oppressions implicated in capitalist difference-making. A cooperatively owned platform wouldn't put an end to algorithmic racism, for instance. This brings us to another important point: sometimes the most emancipatory option isn't to transform how infrastructures are owned and organized, but to dismantle them entirely.

Thinking in Motion

Consider the Stop LAPD Spying Coalition, an alliance that has been organizing against police surveillance in Los Angeles for years. They have successfully pushed the LAPD to abandon two predictive policing programs — programs that led to increased police violence against working-class communities of color. The organizers did not want these programs reformed, but stopped. They were not demanding that the ownership of the algorithmic policing apparatus be "democratized," whatever that might mean, but abolished.

Here is an organization that is taking on tech's tendency to intensify capitalist difference-making, and using the framework of abolition to do so. One can see a similar approach in the emerging movement against facial recognition, as some city governments ban public agencies from using the software. Such campaigns are guided by the belief that certain technologies are too dangerous to exist. They suggest that one solution to what Gandy called the "panoptic sort" is to smash the tools that enable such sorting to take place.

We might call this the Luddite option, and it's an essential component of any democratic future. The historian David F. Noble once wrote about the importance of perceiving technology "in the present tense." He praised the Luddites for this reason: the Luddites destroyed textile machinery in nineteenth-century England because they recognized the threat that it posed to their livelihood. They didn't buy into the gospel of technological progress that instructed them to patiently await a better future; rather, they saw what certain technologies were doing to them in the present tense, and took action to stop them. They weren't against technology in the abstract. They were against the relationships of domination that particular technologies enacted.

By dismantling those technologies, they also dismantled those relationships—and forced the creation of new ones, from below.

Machine-breaking is often a good idea; for more ideas, we can turn to other movements. Tech workers are taking collective action against contracts with the Pentagon and ICE, and demanding an end to gendered discrimination and harassment. Gig workers for platforms like Uber are organizing for better wages, benefits, and working conditions. Within these movements we can find more useful materials to think with, materials that might disclose the contours of a society organized along different lines.

The intellectual is not the only one who thinks. Masses of people in motion also think. And it is the thinking of these two together, in the creativity that results from their continuous interaction, that furnishes the form and content of anything worth calling socialism. This process is messy and circuitous, with many blind alleys and false starts. It involves more time spent moving contradictions around, and creating new ones, than resolving them. But it is the only path to a future where capital's motion finally grinds to a halt, and a different set of considerations—human need, a habitable planet—comes to coordinate our common life. This is how the Left will answer the question of what is to be done, about tech and about everything else: by thinking en masse and thinking in motion, while traversing difficult terrain. ∿

Ben Tarnoff is a founding editor of *Logic*.

Bioplastics Cookbook for Ritual Healing from Petrochemical Landscapes

Tiare Ribeaux

2018 – ongoing

My practice of creating bioplastics involves bio-based alternative materials and hands-on open source methods of making plastic-like materials in order to shift perspectives around the use of petrochemically derived commercial plastics that dominate the landscape of our planet. Using bio-polymers derived from plant or food based materials, and even food waste, I've been experimenting with different types of bioplastics that are compostable and biodegradable.

These materials are transmutative — heated into liquid and poured to dry in a semi-solid form; they shrink, dry, and change shape and color over time. You could say they exist somewhere between the living and non-living. They are remoldable — if broken into smaller pieces and heated in water, they can dissolve and be recast.

Almost 40 percent of commercial petrochemical plastics are made for packaging, with the majority as single-use plastics that go directly into the landfill. Along with other plastic products, these plastics pollute our natural environments, creating the five "great garbage patches" in our oceans made of microplastics, and on land, harming many species and ecosystems. By creating bioplastics, my practice asks: How do we embrace the vitality of matter, particularly in these materials that we are working with, where the question of "life" and "living matter" comes into play, as well as the value of human versus nonhuman life? It also asks how our perceived value of material changes when we are involved in the process of creating them — do they become more precious and harder to dispose of? I've personally kept some bioplastics for over a year. Creating bioplastics is a long and slow process: it is a place of participatory and integrative making outside of our widespread instant-gratification-oriented, ready-to-consume culture. It creates a feeling of preciousness in our most used and disposed-of materials and radically reshapes the relationship we have with them and what we consider "disposable."

My practice of creating bioplastics isn't necessarily offering products or materials that are durable and can be manufactured to scale, but rather I hope to offer provocations, visions, and inspiration for other protocols and methods to be made by others for radically imagining (and creating) a future with alternative bio-based materials replacing single-use or other plastic materials. With our overwhelming dependence on plastic products and packaging, and the detriment to the environment it brings, we need to start to shift our perspectives to imagine material alternatives.

I created an online "Bioplastics Cookbook for Ritual Healing from Petrochemical Landscapes" and have given workshops that envision tactics for reclaiming, rebuilding, and healing from the extractive processes of petrochemical plastics by open sourcing alternative processes that use renewable bio-based materials, urging a collective shift in our material relationships. Using simple ingredients in our kitchens, such as a mixture of agar agar, glycerol, and water — people can become re-engaged with the creation of the materials they use, learn about the ingredients and origins they are sourced from, and the life cycle of these products after their use. My cookbook asks others to think of the process of heating, cooking, and cooling the bioplastics as part of a material ritual, and to think of the intentions they are putting into the bioplastic as they combine the ingredients, and transmute them into new forms of existence. It also asks them to consider other species and mythologies in relationship to materials, their origins, and geographies.

With a background in fashion, biology, digital art, art curation, and production — I've been drawn to creating bioplastics as a method for getting offline, engaging physically with materials, and slowing down. It's also the most accessible "product" of my arts practice — people can right away understand the concept and why it is important in the world — and it is an increasingly relevant topic to discuss. Sometimes I think about every item I touch and if it contains plastic (which is often) and how it might be replaced with something else — perhaps a future bio-material that I may in some way help to shape the existence of. In five to ten years I imagine our packaging being made of something radically different — perhaps out of necessity. Don't get me wrong — I think plastic is an incredible super-material that can be molded into anything. What I don't like is the current relationship we have to it as disposable — how can we value it as much as something like gold? Until then, my personal goal is to continue the conversation, and start making my bioplastic materials into bio-textiles for garments and fashion — it is, at the very least, a fun place to start. ∿

Find the Bioplastics Cookbook for Ritual Healing from Petrochemical Landscapes at:

http://bioplastic-cookbook.schloss-post.com/

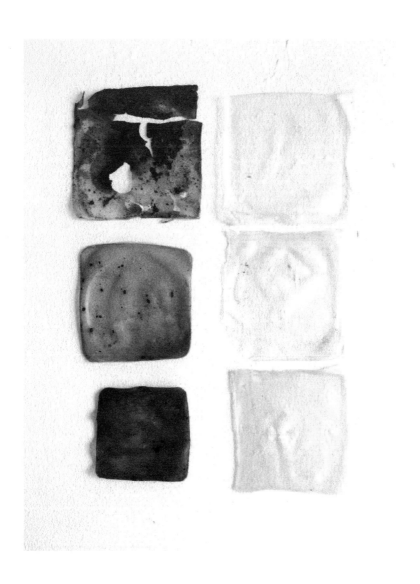

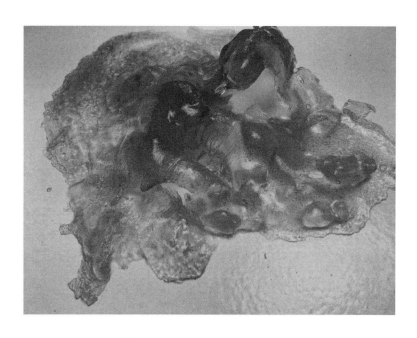

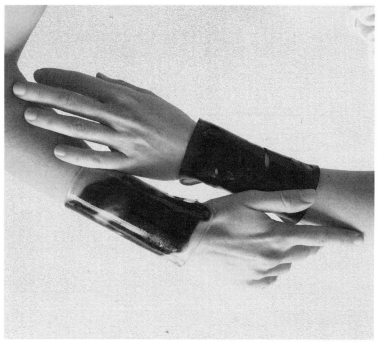

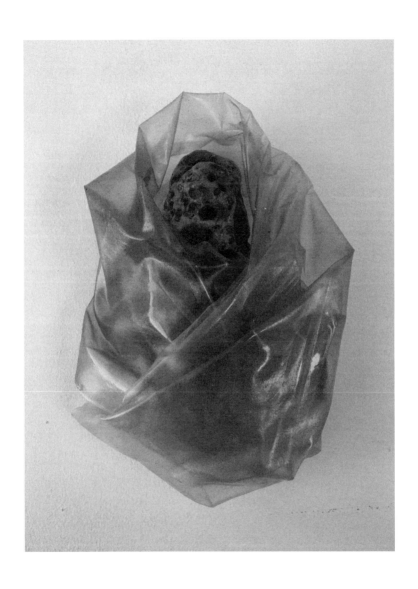

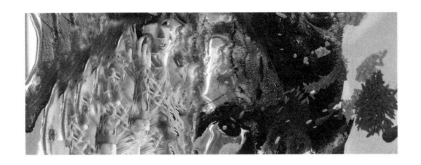

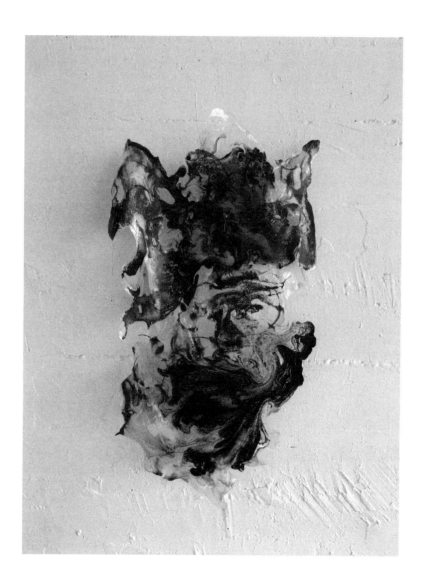

Hack the Planet

Tega Brain on Leaks, Glitches, and Preposterous Futures

While technology critics drag technological solutionism in five thousand words or 280 characters, artist and environmental engineer Tega Brain wields solutions against themselves, in the form of objects and installations. A piece may playfully taunt from the walls of a gallery: "You need me to wear this fitness tracker for your workplace wellness program? Heh, okay." Or another: "You want us to machine learn our way out of the climate crisis? Lol, let's do it." In Brain's work, optimization, datafication, and efficiency are the butt of the joke.

But not all of her work antagonizes the foundational premises of engineering. Sometimes, Brain posits earnest solutions that are "art" because they would never fly within the confines of commercial environmental engineering work, like with her piece that asks: what if, instead of optimizing a washing machine for its ability to clean your underwear, you optimized it for sustaining nonhuman life downstream of your drain?

As humanity reckons with catastrophic climate change, Brain's experiments are raising critical questions about what exactly engineering is for — and what it could be for. We sat down with her to learn more about her work.

———

Though you're currently an artist and professor, you were originally trained as a water engineer. What does that mean?

My first degree was in environmental engineering and I specialized in water, particularly stormwater. This is an area of engineering concerned with how cities, roads, and landscapes are designed to deal with runoff and water flows. It's an area that overlaps with hydrology, which is the study of how water interacts with the environment — how we model and predict rainfall and what it's going to do on the ground. It also overlaps with urban design, because obviously a road is not just a thing for cars to drive on, but also a mechanism to manage water flows in a storm.

"You have to adhere to regulations such as human health standards. But the cost may be the capacity of other, nonhuman species to live and flourish."

Every gutter or drain or pipe is the materialization of a process of data collection, estimation, and modeling and is therefore a wager on environmental variability. How much water might fall? How often? And what is the biggest or smallest rain event the system must be able to deal with? I was involved in designing those sorts of systems.

Towards the end of my time working in that industry in Australia, I was in an area of environmental engineering called "water-sensitive urban design" where we were trying to design *living* systems to improve water quality. Specifically, we were working on improving the composition and quality of stormwater that flowed from housing developments to rivers or oceans downstream. Instead of designing a pollutant trap made out of concrete, we would try to imitate how a natural watershed would work and design a wetland so that stormwater would flow through it and hang out there for a few days before flowing into a creek. In this way, the water would both get filtered and support wildlife and plant growth along the way.

How did you go from doing that engineering work to making art?

There's license in the arts to question very normative assumptions. The engineering approach of designing infrastructures as living systems does still give me hope for how we might rethink human systems more broadly. But I continue to feel quite frustrated with the way that engineering as a discipline tends to frame problems as technical challenges. You're supposed to scope out the political and social forces that are causing an environmental problem, and just slap a technical fix on the end of it. Even the work I was doing—that really nice, innovative, environmental work—was facilitating terrible housing developments full of huge McMansions. It seemed like my job was to make these wildly unsustainable projects just a little bit less bad.

So I started to get more and more interested in different kinds of questions. Like, who and what do we value? What do we think we need in order to have a good life? These weren't questions we asked as engineers, but they were questions I could ask as an artist.

For example, as an engineer, your goal is to minimize risk to humans living in the environment, and to do this, you have to adhere to regulations such as human health standards. But the cost may be the capacity of other, nonhuman species to live and flourish. At some point, you have to think about how you weigh that cost. We urgently need to expand the definition of human health to also include the fates of other life forms. There was very little room in the space I was working in to explore these assumptions and the cost of designing from a solely human-centric perspective.

Unlearning Engineering

"Coin-Operated Wetland" (2011).

Decentering the human feels like a theme throughout your work.

One of my earliest pieces was called "Coin-Operated Wetland." It was an installation that recreated what I was doing as an engineer, but in a gallery space. I built a system where a washing machine was connected to a wetland. The whole installation was a closed system, where the water that was used to wash clothes ultimately ended up in the wetland and then circulated back to the washing machine. What if you show people that there is no downstream? What if you're confronted with the life forms that are directly impacted by your actions?

The night before that show opened, I was incredibly stressed because working with water is so hard! With software-based work, if you get a glitch, you edit the code and even use the mistake to inform the aesthetic of the work. But mistakes in a system involving water can result in, you know, flooding.

Also, in order for a living water treatment system to work, you can't use disinfectants like chlorine because it will kill all the bacteria and plants. So from a human health perspective, that system didn't actually comply with health standards and I kept thinking, "Oh my God, someone's going to put their hand in it and put their hand in their mouth, and I'm going to get sued." That was the engineer in me, thinking about risk minimization.

In the end, the system was actually pretty successful; the plants were happy and, even though the laundry water wasn't treated to drinking water standards, the T-shirts we put in the laundromat came out looking clean. Of course, it required more maintenance and labor than a washing machine you'd have in your house, and it was less efficient by some measures; we could only do one load per day because that's the pace at which the plants could consume the water. But if we're going to shift away from seeing ecosystems strictly as service providers and towards a more negotiated, reciprocal relationship with them, our systems are going to need a little more give. That project was about seeking a balance, and exploring how to build infrastructures that are not optimized for humans alone.

I mean, you'd never be allowed to build something like that as an engineer. The client would sue you.

That balancing act reminds me of something engineer and professor Deb Chachra wrote in one of her newsletters. She wrote, "Sustainability always looks like underutilization when compared to resource extraction."

That's beautiful. Deb also writes about infrastructure as being care at scale, which I think is a nice way to think about it. Could there be a model where infrastructures don't just care for humans, but also care for the ecosystems where they're acting?

I'm obsessed with water leaks for that reason. If you look at a water pipe at the point where it's leaking, you usually have these little gardens popping up, all these little ecosystems that are taking advantage of the water supply. There's been fascinating research published on how leaks from water distribution systems in cities actually recharge groundwater aquifers because most of these systems leak 10 to 30 percent of their water.

Of course, there's also research going on at MIT and all these engineering schools on how to develop little autonomous robots that go into the pipes and find the leaks and plug them up. From the perspective of design and engineering, the system is not supposed to be porous; leaks are a problem, an inefficiency. But it actually takes more than just humans to make the city. What about the street trees that depend on those leaks? So then the question becomes: is there a way we can share resources with other species rather than completely monopolizing them?

The shift from looking at unintended side effects, of leaks for example, to intentionally surfacing or creating side effects reminds me of your project "Unfit Bits." You and your collaborator Surya Mattu demonstrate all these ways to "hack" a Fitbit by making it register steps when you're not actually taking them, such as attaching the Fitbit to a dog's collar or the wheel of a bicycle. That project is about deliberate subversion — side effects intended by the hacker, but not by the creator of the system being hacked.

"Unfit Bits" (2015)

I haven't really thought about leaks and "Unfit Bits" together, but that project is also about the manipulation of infrastructure and the deliberate glitching of a dataset for a different political end. Whether you look at leaks in a water system or leaks in an information system like the Panama Papers or the Snowden leaks, both are about a redistribution of power.

In the context of fitness trackers and employer-provided health and life insurance, the pitch is that tracker data provides an accurate picture of a person's health. That's a very political claim, especially in the US where access to insurance is commodified and not universal. As a result, employers are handing over employee fitness data to private insurance companies as part of workplace wellness programs, and some of them even penalize their employees for refusing to wear these trackers. Life insurance companies are using Fitbit data to help determine premiums. You've got this very fraught situation where a dataset is playing a critical role in how people get access to essential services.

With "Unfit Bits," you author your own dataset and optimize it for what you want rather than being subject to what it says is the objective reality of your life. If you want to see how a system works, look at what the system deems inaccurate or inefficient. Pay attention to what's called an error. In a water system, it's a leak. With a Fitbit, it's anything that doesn't meet the narrow definition of a step. On the flip side, many gestures that aren't steps get measured as steps. "Unfit Bits" exploits that.

With both leaks and glitches, you're poking holes in the idea that systems are perfectly closed and objective.

Yes, and it's also me trying to unlearn an engineering worldview. When you're trained as an engineer, you're taught that you're going to make a system that solves a problem. Very rarely do you get to the point of asking: is the problem we're solving for the same for everyone? Who gets to decide what qualifies as a problem and what are the tradeoffs in how it is defined? There are these universal ideas of what's efficient. Well, efficient for whom?

Not looking at the world as an engineer is also about embracing inefficiencies or using them to to tease out what we call success in a system.

Simulation Machines

Part of what makes "Unfit Bits" work is the absurdity of putting a Fitbit on a dog collar to convince the device that you ran five miles. Your project "Asunder" also uses absurdity, but to make a point about optimizing the environment. Tell me about that project.

It's basically a simulation machine. Julian Oliver, Bengt Sjölén, and I built a computing system to generate fictional

"Asunder" (2019)

geoengineering proposals. The work cycles through publicly available satellite imagery of the world—tiles taken from the Landsat 5, 7 and 8 satellites (6 got lost). From that dataset, we chose a series of sites that have gone through significant change over the last thirty to forty years and we show these places using the historic satellite tiles for that site.

From there, our computer generates preposterous scenarios for geoengineering that site, like rerouting a river or recombining the site with another site from across the world. It generates lots of scenarios using a GAN—a machine learning method that uses existing images to generate new images—to stitch the satellite tiles together. You end up with these surreal, dreamlike Landsat tiles that are made up of edited landscapes. Then the system chooses one possibility, analyzes the land use changes in it, and inputs that data into a climate model to estimate how the change would impact the environmental performance of the earth overall.

One of our sites is Silicon Valley. The system generated a geo-engineering scenario where it took a lithium mining region in Chile and transplanted it into Silicon Valley. In another scenario, a region in Antarctica is recombined with an area of industrial agriculture in the US Midwest. Using the climate model, the simulation then gradually tries to estimate what that would mean, climate-wise.

The project takes the solutionism that you hear in geoengineering spaces to the most ridiculous extreme possible. These solutions are totally not viable.

I know it's generating ridiculous scenarios, but I can also imagine the EPA or NASA wanting to do exactly this.

Yeah, and if you look at climate forecasts a hundred years out, there's a lot of extremely bizarre land use changes that are predicted unless we can drastically change our systems of production. An ice-free Arctic. Agricultural areas shifting to higher latitudes as frozen areas in Russia and Canada thaw. The complete desertification of agricultural areas that are viable today. And if that's not bad enough, all of the southern wine-growing regions in Europe are predicted to become too dry to support wine production anymore.

This is the information that's coming from the scientific community today. So although the system we built with "Asunder" generates scenarios that feel preposterous, we live at a moment where scientists are predicting an even more catastrophic future.

We're all trying to assimilate that view of our future. I think anybody who's done even a little bit of reading on the subject must feel a deep sense of dislocation. The narratives we grew up with around modernity and technology and progress are really at odds with what the science is telling us.

How did you all decide on satellite imagery and machine learning as the mediums for this piece?

"The system generated a geoengineering scenario where it took a lithium mining region in Chile and transplanted it into Silicon Valley. Then, it gradually tries to estimate what that would mean, climate-wise."

They're part of a long tradition. The history of weather prediction is entwined with the history of computation: the first electronic computer, the ENIAC, was a military technology developed in the 1940s, and then put to civilian use. One of the use cases was weather prediction; in the 1950s John von Neumann developed the first weather prediction techniques on the ENIAC. So this is a long-running historic project that has produced the knowledge of how we've changed the composition of the atmosphere and what that means. We wouldn't be able to understand climate change if we didn't have these technologies. They have given us a view of the world that would otherwise be impossible.

At the same time, these simulations of the world can make you forget that there are other ways to know. Computing is so seductive that way. It makes you forget that there's always something the simulation can't capture. And it turns out that all these decisions have to be made about *how* the simulation or the model is built, and those all impact the end result.

"Asunder" is also connected with work I've been doing in the past year or so around machinic perspectives of environmental systems and what looking at the environment as a computer does. How does it make us see the biosphere, and how does it produce and foreclose certain possibilities?

These questions are really important, because we have some urgent environmental challenges to deal with. And this is happening at the exact moment where we also have a surplus of computation. Everyone's like, "What can we do with all this computation? I know! We can solve climate change and extinction and whatever!"

You've been making work about the environment, nonhumans, and environmental timescales for a decade. My sense is that the language of "climate crisis" and "extinction events" has recently become mainstream in a way that it wasn't before. What has it been like to watch that happen? Has it changed how your work is received?

It's a huge relief to see these issues become more widely discussed and to see organizing and activist work happening in the mainstream. We need more of that, and we need our collective action to be more extreme because the political class has decided to leverage denial for their economic gain. There's no question that we need to be doing everything we can to take power away from those people. It's horrifying that denial and doubt are being used very strategically by powerful people not only in the US, but also in Australia where I'm from.

I'd love to see much more experimentation around how to reconfigure relations and trouble the human-centeredness of technologies and infrastructures. I'm excited to see more work that takes up what it means to attempt to *optimize* an environment. So often, we think about data in service of prediction and control as the primary way to encounter an environment. And yet there are so many other ways to know a place. ∿

Tag Yourself

by Sara Stoudt

As environmental research budgets get slashed, can amateurs fill in the gaps?

————

Since 1992, volunteers have tagged more than 1.5 million monarch butterflies. Tagging a butterfly involves capturing it in a net, attaching a label to its wing, and releasing it back into its habitat. The identifying information on the tag goes into a database that tracks the monarch's famous migrations to and from California and central Mexico. Chip Taylor, founder and director of Monarch Watch, the nonprofit that organizes this vast volunteer effort, says the process is easier than it sounds and that the monarch butterfly is hardier than it appears. Even students can participate, and they often do so as part of their science classes, guided by teachers. Taylor noted that 2018 held the record for the largest number of butterfly tags distributed, with over 320,000 mailed to interested volunteers across North America.

Monarch butterflies are unique. Not only are they resilient to being captured and tagged by volunteers, they are also widely loved. In addition to Monarch Watch, there are numerous Facebook groups dedicated to the orange and black beauties, including Monarch Butterfly Garden with over 50,000 followers, and The Beautiful Monarch with over 25,000 members. Volunteers are eager to study and protect them.

What about species that aren't as charming? Who will account for the creepy crawlies and the drab species? These are some of the questions underlying the scientific community's increasing reliance on crowdsourced environmental data. Monarch Watch helped pioneer the crowdsourcing model, but its analog, low-tech approach has since been overtaken by a wave of apps and platforms that have made the process of collecting data much more accessible. The proliferation of smartphones and the rise of the mobile web has enabled more people to contribute more observations, on more species. And while this phenomenon has been growing for years, it has acquired a new urgency since the election of Donald Trump. Under Trump's leadership, scientists have seen a decrease in funding for environmental research and an official denial of climate change despite mounting evidence. This means that research scientists need crowdsourced data more than ever before, incomplete as it may be.

"Community scientists" can help. In addition to not relying on government budgets, these nature-loving, albeit untrained and unpaid, members of the public have another advantage: they can use apps to collect data about more species, over a larger physical area, than the comparatively small number of professional environmental scientists can. But if community scientists completely drive scientific research agendas, society risks losing valuable information about critically important species. Community science efforts can only augment scientific research, not replace it.

Shrooms at Scale

The aughts saw an explosion of community science apps and websites that let users collect photos, dates, times, and geo-coordinates for different plant and animal sightings. Love birds? Join the hundreds of thousands of users on eBird, a database started in 2002 that now has hundreds of millions of bird observations. More of a mushroom person? There's Mushroom Observer, which has spawned hundreds of thousands of observations from thousands of participants since it started in 2006.

"Community science efforts can only augment scientific research, not replace it."

There are online spaces for generalists as well. One example is iNaturalist, an app, website, and online community that launched in 2008. A user on a hike can upload a photo of a plant or animal and have immediate access to the platform's community of naturalists to help identify it. iNaturalist has amassed 25 million observations, over 10 million of which are research-grade. The platform is also a popular public engagement tool for museums and other institutions that use it as part of their programming.

The scale of data collection that's possible with these platforms surpasses anything a team of scientists could ever hope to match over the course of a career, even with ample funding. Users of iNaturalist and eBird have collectively recorded observations of over 200,000 species.

This data often makes its way into scientific research. Despite Mushroom Observer being dominated by amateurs, one of the site's developers, Joe Cohen, says that trained researchers actively participate in forum discussions, track particular species, and share specimens with other users. iNaturalist, Mushroom Observer, eBird, and Monarch Watch all submit user-collected data to the Global Biodiversity Information Facility (GBIF), a central repository that brings together data on species occurrences from tens of thousands of different data sources. Scientists can easily access the data filtered by their species of interest. Data collected by app users and accessed via GBIF has been cited hundreds of times in scientific research.

Burnt Out on Butterflies

But the apps and websites that make this large-scale data collection possible are not designed for conducting scientific research. iNaturalist, for example, makes clear that its first priority is "to connect people to nature"—the breadth and volume of the environmental data collected is a fortuitous side effect of community-building. And since producing data specifically for scientific research is not what these platforms are for, sampling problems abound.

Data collection sites that are near users' homes or are easy to get to become hotspots for observations, regardless of their value for scientific research. While scientists often travel to field sites to collect data, Mushroom Observers, for example, typically collect data near where they live.

When community scientists do travel, they may be more interested in going to places where they can expect to find a particular species, either because the species is more prevalent or because the place supports data collection in some way. Chip Taylor of Monarch Watch remarked that monarchs are better represented in Iowa because its county conservation boards facilitate tagging efforts. Volunteers also prefer to report species they find interesting. Rare species may be overreported because people are excited to see them and may even travel specifically to see them—a data collection pattern encouraged by some platforms' design features, like eBird's rare bird alerts. In 2018, the monarch butterfly was the most observed species in iNaturalist's research-grade observations in ten of the forty-eight states in the continental US, though it's unlikely that monarch butterflies are the most prevalent species in any of those states. Meanwhile, relatively little data was collected by users on the less charismatic *Bridgeoporus nobilissimus* mushroom.

> *"Since producing data specifically for scientific research is not the priority of community science platforms, sampling problems abound."*

Misidentifications can also be a problem, even though companion apps for community scientists generally require multiple identifications before an observation is confirmed, and some apps like iNaturalist use computer vision to suggest identifications. In an effort to ensure the quality of their dataset, scientists using crowdsourced observations for research may treat the number of contributions a user has made as a proxy for data quality, and filter out users with a weak contribution history.

Since the charisma, visibility, rareness, and location of a given species can all affect data collection in ways that don't necessarily reflect the species' actual distribution, it can be difficult to determine whether an absence of observations corresponds to a real decline of the species or something else. Some platforms try to account for this in different ways, but others, wary of user attrition, are hesitant to add barriers to submitting observations. After all, what matters most to the platforms is attracting and retaining users. Helping out scientists is a secondary concern.

"*Gaps in data take infrastructure and resources to fill.*"

Bridging the Gap

Researchers do their best to account for the limitations of crowdsourced data. They add instructions to particular data collection efforts or work in tandem with volunteer data collectors on training initiatives about the need for high-quality data. "Data fusion" methods and integrated population models have also become popular tools to bring data together from different sources, weighing the strengths and weaknesses of each dataset. Gaps in community science data can inform scientists' future data collection, providing opportunities to improve sampling design and data collection efficiency.

Despite their limitations, then, platforms like iNaturalist, Mushroom Observer, and eBird are still valuable for scientists. The scale of biodiversity is such that scientists alone cannot record everything, particularly in an era of slashed research budgets and anti-science public policy.

Still, defunding science has serious consequences and we can't afford to narrow our focus at such a critical moment of ecological change. There are inevitable, irreparable gaps in data collected by community members. Going back in time to collect better data on a particular species or in a specific region is impossible. When scientists are more reliant on data collected by volunteers, the fluctuating interests of the public can destabilize research efforts. We still need data about boring species, and from faraway places. These gaps take infrastructure and resources to fill, and we ignore them at our peril. ∿

Sara Stoudt is a PhD candidate in statistics at the University of California, Berkeley and a Berkeley Institute for Data Science Fellow.

Water is Life

Nick Estes on Indigenous Technologies

Water is life. Not in a mystical or romantic way, but in the material way that all humans and countless nonhumans need water in order to stay alive. From August 2016 to February 2017, thousands of Native and non-Native people gathered at Standing Rock to fight for a world structured around a central tenet of Oceti Sakowin philosophy: we want to live and we want our children to live, so we have to protect the water. The opposing philosophy, enforced by multiple state police departments, private security contractors, and the US Army Corps of Engineers: profit is our birthright and we will extract it by any means necessary.

This struggle has been ongoing for generations. In his 2019 book *Our History is the Future*, Nick Estes, a citizen of the Lower Brule Sioux Tribe, situates the months-long encampment at Standing Rock within centuries-long traditions of Indigenous internationalist resistance to white supremacist imperialism, settler colonialism, and capitalism. Estes is a professor at the University of New Mexico and an organizer with The Red Nation, the Indigenous resistance organization he cofounded in 2014.

We sat down with him to talk about narratives of technological and scientific progress, the Red Deal, and the problem with land acknowledgements.

———

I wanted to start with your first day at camp at Standing Rock. In your book, you write about digging compost holes with an Ojibwe relative and building a kitchen shack with a Palestinian network admin. It seemed like an incredible logistical feat that brought together people from all over. Can you talk about the infrastructure you all built there and what made that convergence possible?

My first day at camp was late August 2016, before the dog attacks. We arrived to bring supplies, and we set up camp for about a week. Some of us from our organization, The Red Nation, had to leave, but some of us stayed for a long time. One of our people stayed until the last day when camp was evicted in February 2017.

By and large, the infrastructure of the camp was organized around tribal nations. Our tribe, the Kul Wicasa, or Lower Brule Sioux Tribe, set up our own camp. Next to us was the Ihanktonwan, or Yankton, and next to them was the Oglalas. Then there was the Cheyenne River Sioux camp and then across the Cannonball River, there was the Rosebud Sioux camp. The camp structure took on an organic shape. Later on, other organizations and tribal nations filled in.

Because of the culture of Native people in general, our camping and outdoor life is really well organized. We have a depth of communal knowledge about those subjects. Even though we are colonized and confined to reservations and don't live the life that we once lived, we still have a seasonal cycle of migration and gathering. Summers are very community-oriented and organized around a kind of camp life, whether it's Powwows or fairs or Sun Dances or whatever. Then in the winter, we go back to our more settled homes. Camp life at Standing Rock reflected that.

Everything was organized around need, so the first thing that went up were the porta potties. Then came the kitchens, followed by the donation tents where people could get camp supplies they didn't have. It reflected the traditions of Indigenous people: if you didn't have enough, you were still taken care of. Many people see Indigenous generosity as a weakness, but it's one of our strengths.

Over a longer period of time, people developed internal political processes, both formal and informal. Not everyone was Lakota or Indigenous, and with that many people sharing space, there had to be some kind of community agreements. There were community councils where non-Indigenous people had a say. The camp infrastructure wasn't meant to be permanent, but it suited the purpose.

Was there power or Wi-Fi?

No. Or maybe there was for a moment. There was a place called Facebook Hill, which was the only place where you could get good cell phone reception. You would see people up there checking the internet, broadcasting to Facebook or checking email.

I ask because the conventional narrative about other large mobilizations like Occupy or Arab Spring tends to emphasize the role of social media. How do you think about technology, whether within the context of the encampment at Standing Rock or more broadly?

Technology is interesting because its value is socially constructed. For Native nations, technological progress is usually top-down. It's usually something that's forced on us. More generally, capitalism as a social process has devastated our communities. It has ensured that we don't have self-determining authority over the means of production that are located on our land. Take the Navajo Nation: there are all these fracking rigs going up there. The road systems are created as infrastructure for fracking rigs. They're not infrastructure for the people who live on the land.

My friend, the poet Mark Tilsen, made a joke when we were discussing what the future would look like. I said, "Indigenous peoples aren't protesting the construction of wind turbines and solar panels on their land." And he replied, "*Yet.*" It's true: regardless of what the technology is, who has the power to decide how it will be implemented and managed? Who will shoulder the burden of the transition away from fossil fuels? Take those electric cars that run on batteries made from rare earth elements—those elements have to come from *somewhere*. Those wind turbines have to be built on *someone's* land.

At the same time, I would say that Indigenous ontologies and ways of being are social systems that value different things than settler ontologies, so our technologies look different. Indigenous technology gets cast as primitive, like it may have been useful in the past but no longer has any relevance. But that's not true. Assembling communal life is in itself a technology.

That dynamic, where technology by default means settler technology or capitalist technology as opposed to Indigenous technology, also operates with the law.

Your book shows that US law is not "the" law but is "settler law," one of many possible legal frameworks. And one subtext of the book is that settler law is a technology for dispossessing Native nations of their land and replacing Indigenous people with settlers and infrastructure to support settler life. There's a Lakota concept for this that you describe: *Woope Wasicu*, or "law of the colonizer." Can you talk more about that?

These are ways that we're racialized: we're constructed as not developing socially valuable technologies and we're constructed as lawless — not having forms of order, or having forms of order that are not legible to the settler state. Erasure is a social technology that makes the taking of our land much easier. It's done not just at the level of the imagination, but enacted through the law itself.

> *"The road systems are created as infrastructure for fracking rigs. They're not infrastructure for the people who live on the land."*

In my book, I mention this 1823 Supreme Court decision that said that Indigenous people only had *occupancy* rights to our land — not full title — so settlers who "discovered" our land could legally take it. That ruling was based on the "doctrine of discovery." The Chief Justice in that case, John Marshall, cited a fifteenth-century papal bull called the "Doctrine of Christian Discovery" that was used to legally justify Portugal's claims to land in West Africa. The reasoning was that, just like non-Christians in West Africa were considered "savages" who couldn't own their own land, we couldn't have full title to our land because we're not full humans who exist at the level of civilization — a doctrine decided and standardized, of course, by the colonizing nation.

So settler law comes in to impose itself on ours. And what's interesting about US settler law is that even though the United States claims to be a democratic republic, it has a very covenant-based government, meaning that it derives its authority from a constitution that hasn't changed much since it was written. The United States is different from other liberal capitalist democracies in that it came into existence as a capitalist democracy from the get-go. It didn't evolve from feudalism like democracies did in Europe; it supplanted itself on top of what already existed, in order to destroy what already existed.

Consequently, US settlers were one of the first nationalities to define themselves against the people whose labor and resources they depended on, whether it was African slaves or Indigenous land. There are core principles of American identity that revolve around white supremacy, land ownership, xenophobia, anti-Indigenousness, and anti-Blackness. Those principles are ingrained not just in the Constitution, but into the broader social fabric of the United States.

Those principles also helped compose a highly centralized national identity. In response, the multitude of disparate Native nations became centralized into fewer, more unified identities. Our Indianness as a universal identity that we share is always defined against what we are not, and what we are not is a colonizing nation.

Another thread that runs through your book is the extent to which feats of engineering and scientific "progress" literally

come at the expense of Native life and land. You write about the Pick-Sloan Plan, which was sold to the non-Native public as an innovative hydroelectric power project. Can you talk more about that?

In the name of providing cheap hydroelectricity to settlers and making the prairie bloom through irrigation, the Pick-Sloan Plan called for the construction of five dams along the Missouri River. So between 1946 and 1966, the US Army Corps of Engineers condemned and seized 550 square miles of Native land through eminent domain. The dams also flooded seven Lakota and Dakota reservations and forced thousands of people to relocate from land they had lived on for generations.

Those dams were imposed on us by the US military. Hydroelectric dams have a lot in common with nuclear power plants in terms of how they're centrally and hierarchically managed, how they produce power, and how they're ingrained within the military-industrial complex. Hydroelectricity and nuclear energy also both get lumped in as "green" technologies, but I would contrast the impact and management of those particular forms of technology with solar grids and wind turbines, which are very decentralized.

If you turn off the hydroelectric dam, the impact is catastrophic. The same goes for a nuclear power plant: if you're not cooling your nuclear rods, there are disastrous downstream—literally, down the stream—consequences. You need hierarchical management built in, to keep people safe. But the existence of those threats is a manmade crisis that naturalizes and justifies that hierarchy once it's been created.

On the flip side, solar power and wind power are decentralized. You knock a couple wind turbines off the grid and it doesn't have any effect. Those are some of the things I think about. And even the fossil fuel industry is thinking about this. They want to recentralize those decentralized green technologies. It's like the internet: everybody thought it was going to democratize everything, and now it's been totally privatized and commodified. That's something we have to fight in this energy transition.

Camp Tech

It's hard to talk about the internet without talking about centralization, and also without talking about surveillance. One of the metaphors we have for online surveillance is the panopticon. But the scholar Simone Browne makes the case in her book *Dark Matters* that the origins of surveillance also lie in the slave ship and the forms of racialized policing that emerged from the plantation. Would you add the reservation to this set of ways of thinking about surveillance?

Yeah. I don't think it's a competing framework. I think it's complementary. I've been thinking about camps as a technology of surveillance and control, and I would consider reservations to be a kind of camp. In the book, I talk about a few types of camps. There's the resistance camp — the blockade — which has long been a tactic of Indigenous people. You saw it from the late 1960s through the early 1980s with the occupations of Alcatraz and Wounded Knee, and the Yellow Thunder camp in the Black Hills. You also saw it more recently with Standing Rock and the Unist'ot'en camp in British Columbia. Mauna Kea, where thousands of people are camping to protect a sacred Native site from a billion-dollar telescope, is becoming a resistance camp. We could go on and on.

But there's also the concentration camp. The US concentration camp originated as a technology of control specifically for Indigenous people. Under Abraham Lincoln, thousands of Dakotas were put in a concentration camp at Fort Snelling. In the Southwest in the same period, Navajos were subject to forced marches and imprisonment in camps. And, of course, reservations were and are concentration camps. Russell Means of the American Indian Movement once said, "Pine Ridge is concentration camp #334." On our tribal IDs, we each have an assigned number that corresponds to our reservation. I'm from Lower Brule so mine is 343. The concentration camp evolved into the apartheid Bantustan system in South Africa. The architects of that system were looking at the reservation system in Canada because they shared an affiliation with the British Crown. So these technologies are co-constitutive.

The last type of camp are what Indigenous activists have called "man camps." These begin as transient settlements of extractive industry workers who set up in an area temporarily and then leave. They're the shock troops of capitalism. But what starts as temporary extractive infrastructure eventually becomes permanent outposts.

Could you give an example?

My hometown, Chamberlain, South Dakota, used to be called Fort Kiowa. And it was a trading fort, a militarized encampment of primarily men who were killing tons of animals to extract furs. Now, it's a racist border town. Many people think of the US-Mexico border when they think of border towns, but here I'm talking about the white-dominated settlements bordering Indian reservations that were once man camps and have now become permanent fixtures.

The penetration of capitalism into non-capitalist economies is always accompanied by extreme violence, and it's not just a process that starts and then ends—it's ongoing. There was a time when US hunters and soldiers would go out and kill as many buffalo as they could. That was it; that was the whole goal. Miners did the same thing. And now you have man camps of oil workers who go into a region, run the oil extraction machinery, and then leave. These man camps obviously exist to extract resources—whether it's animals or people or minerals—but they also perform a certain social function: they reorder societies toward the accumulation of capital.

You see this now in the West Bank. Israeli settlers occupy these outposts in order to reshape the landscape and disrupt the social world of the people who were already there. The state deploys the Israeli Defense Forces to protect the settlers and put down Palestinian resistance. Sometimes there are more soldiers than settlers: they're just there to protect a small sliver of illegal settlement.

So those are three technologies of surveillance that we can think about: an Indigenous countersurveillance program of creating resistance camps, the state-sanctioned concentration camp, and the public-private partnership that is the man camp.

When I asked the surveillance question, I thought you were going to talk about racist policing around reservations. But it sounds like the policing of Native people also happens in a more informal way through these man camps that pop up around extractive industries on or near Native land. So is the police and prison abolition work you do with The Red Nation also about fighting those extractive industries?

The Missing and Murdered Indigenous Women (MMIW) issue is a perfect way to talk about carceral abolition work in that context. The activists who developed the MMIW framework were connecting it explicitly to the extractive industries. But there's also a framing that says the problem is due to a lack of law enforcement as opposed to understanding police as part of the problem. If you read the reports on the MMIW epidemic, the perceptions and actions of law enforcement confirm that, as an institution, the police perpetuate the problem of violence against Indigenous women.

"The penetration of capitalism into non-capitalist economies is always accompanied by extreme violence, and it's not just a process that starts and then ends — it's ongoing."

New Mexico has the highest number of MMIW cases in the US. What conditions contribute to that? New Mexico is undergoing an oil boom. Where are these women disappearing? They're disappearing in border towns. Gallup, New Mexico was once a coal-mining community. Santa Fe was once a place where they bought and sold Native slaves. The same with Albuquerque, which is where I live. These are now permanent border towns where there are high rates of violence against Native women, as well as against LGBTQ and Two-Spirit people [*Eds.: This is a Native term for gender-nonconforming people, distinct from the non-Native concept of LGBTQ identity.*] In regions where oil and gas is taking off, you see this increased violence. Unfortunately, instead of throwing out the extractive industries, the solution so far has been to go to the police.

Let's bring the surveillance conversation back to Standing Rock. In your book, you write about emails and other documents that came out after camp was evicted, where police and security contractors were discussing counterinsurgency tactics to use against Water Protectors. They were talking about "riot control agents," aerial surveillance, and infiltrating camp. Did people at camp know this was happening?

There's a naive trust in Facebook, Twitter, our cell phones—all these things we're socialized to use day-to-day and bring into our intimate lives. Even technologies that are supposed to be encrypted were hacked. We don't know how they were hacked, but information was getting out.

> "We learned after the protest that the Sante Fe Police Department had issued a sealed warrant and that Facebook had turned over all of our communications on Facebook Messenger."

I think that naive trust is partly because there's a generational gap between movements today and those of the past that experienced the violence of COINTELPRO, the FBI's counterintelligence program that targeted the Black freedom movement, the American Indian Movement, and the antiwar movement in the 1960s and 1970s. We don't believe we are under constant surveillance, even though there has never been a point in human history where we are under such constant surveillance.

So those documents weren't surprising. What we can take from them is how police officers and private security firms like TigerSwan were and are connecting different struggles. We're often taught to silo our struggles, to say, "This is a Black issue; that's a Native issue; that's a Palestinian issue." But they see it all as one, and we should too. North Dakota state police shared a federal "Field Force Operations" manual that references Ferguson. They are drawing connections among many different issues: violence against migrants crossing the border; the policing, criminalization, and surveillance of Water Protectors at Standing Rock; the dehumanization of Black people in Ferguson and Baltimore. Our adversaries see themselves as participating in a global counterinsurgency war, and we can't underestimate the power of that alliance. It's not a secret that Facebook works hand-in-hand with law enforcement.

Our group, The Red Nation, dealt with this when we were planning a protest against the Entrada, a celebration of Spanish reconquest after the 1680 Pueblo Revolt that ultimately ended up being abolished by the city. We learned after the protest that the Sante Fe Police Department (SFPD) had issued a sealed warrant — so we never saw it at the time — and that Facebook had turned over all of our communications on Facebook Messenger. As a result, on the day of the action, the SFPD brought in twelve different law enforcement jurisdictions. There was a huge police presence with sniper nests and everything. Eight of our people got arrested. We didn't find out until later that they had access to our Facebook messages. Most of what was on there was irreverent memes about the cops, but that was a wakeup call. We no longer write anything on our phones or social media that we're not willing to share in public, no matter how private we think the conversation is.

The Red Deal

I want to change gears to talk about the Red Deal, The Red Nation's proposal for climate justice and decolonization. What would you say are the main pillars?

Our program is influenced by the divest/reinvest strategies of Standing Rock and the Movement for Black Lives. At Standing Rock, Water Protectors called for divesting from fossil fuel industries. The Movement for Black Lives platform calls for divesting from carceral institutions and reinvesting in the things that people need to live—instead of the things that put us in jail.

The Red Deal focuses on the state itself as opposed to industry because it's the state that keeps the extractive industries intact. Who else was at the pipeline protests? The police. What allows the criminalization of Native people? The carceral legal apparatus. What prevents colonized nations from throwing off the yoke of US dominance so they can develop? The US military. So demilitarization and carceral abolition are two main pillars of this program. We estimate that divesting from those state institutions would free up about a trillion dollars to reinvest in things like hospitals and healthcare and land that has been destroyed here, as well as in other countries that have been damaged by the US military.

We're also using the idea of Alexandria Ocasio-Cortez and Ed Markey's Green New Deal, which essentially argues in its legislative text that every social justice issue should become a climate justice issue. Indigenous people have long been the most confrontational arm of the environmental justice movement, but have received the least attention when it comes to actually making policy. The Red Deal says that if we're going to imagine carbon-free economies and the end of fossil fuels, then we also have to talk about decolonization. How are we going to build wind turbines but not give the land back to Indigenous people?

The Red Deal stands for a caretaking economy. If soldiers and the police are caretakers of violence, then we need to contrast those value systems with people who are caretakers of human and nonhuman life. That includes teachers, nurses, counselors, mental health experts. It also includes land defenders and Water Protectors.

We all need water and land and forests to live. But when you walk into a restaurant, who gets a discount? Military and police, who, by the way, tend to be men. That reflects a value system. Caretakers tend to be women, and caretakers of the land tend to be Indigenous. If we look at the anti-protest and anti-BDS laws *[Eds.: the Boycott, Divestment, and Sanction movement is a Palestinian-led campaign "to end international support for Israel's oppression of Palestinians"]* that have gone through state governments, they criminalize caretakers. So that's what we mean when we talk about investing in a caretaking economy that seeks to live in a correct relation with each other as human beings and nations, as well as a correct relation with the nonhuman world.

On the topic of lip service, what do you think of land acknowledgements? I recently came across one on the website of Sidewalk Labs, an Alphabet subsidiary that's spending $900 million to build a "smart city" on the Toronto waterfront.

Here's what I think about land acknowledgements. I ride a bike to work. Imagine I wake up one day and my bike is gone. I'm late for work. Maybe I'm going to get fired and I won't be able to feed my family, but I'm shit out of luck. And then some guy rolls by on my bike and is like, "Hey. I want to acknowledge that I'm riding your bike. I know it's really bad that I stole it, but I hope we can work towards reconciliation."

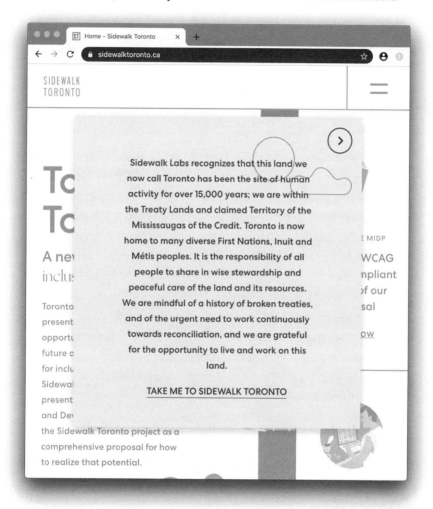

And then he cruises away on my bike!

I guess it's not that land acknowledgements shouldn't happen, but they just make me think, "Uh, okay. Great."

But this kind of gesture isn't limited to land acknowledgements. People often use the language of decolonization when they're not actually talking about giving back our land. What they're actually talking about is *indigenization*. That's an inclusion framework: let's include Indigenous people in their own dispossession. Let's have more Indigenous clergy! Let's have more Indigenous people in the police, in the military, in the forces that occupy our land!

Good Relations

In the vein of concrete questions about Red Deal implementation, I want to ask you about tradeoffs that tribal governments have made over the years. In the 1960s, Fairchild Semiconductor, a microchip manufacturer that was one of the first major firms in Silicon Valley, built a factory on a Navajo reservation in Shiprock, New Mexico with the support of the Navajo Tribal Council.

The one where the workers went on strike for better working conditions and then Fairchild shut it down?

Yes. Presumably, the Navajo Tribal Council wanted to bring jobs to the reservation. You've written about the tradeoffs that tribal governments navigate in deciding whether to participate in coal production or chip production or even solar power production on Native land. Do you think it's fair to talk about that within the context of a Red Deal, or do you think the question is more about why tribal governments even have to think about those tradeoffs?

There are a lot of examples of this. There's a Raytheon facility right outside of Farmington, New Mexico on Navajo land, where the workforce is 90 percent Navajo. That facility makes the microchips for the Israel's Iron Dome missile defense system. The Alaska Native Corporation employs a lot of the private security forces who work at the child and family detention centers on the US-Mexico border. They also have contracts building what are essentially military bases in the Pacific. The Cherokee Nation has contracts to build State Department facilities in the Green Zone in Baghdad. There's also a federal law that gives preference to Native businesses for lucrative defense contracts. These are the opportunities we get, and we have to take them because our subsistence economies have been annihilated.

People say the Navajo Nation is dependent on coal and oil and gas, but I would actually say that the Southwest is dependent on the Navajo Nation producing coal and oil and gas because no one else wants to do it. No one else will have the generating station on their land because it's one of the dirtiest coal-fired power plants in the country. So Navajo lands have been sacrificed—whether it's been for coal, oil, and gas, or something like uranium. The same is true of Pueblo lands: the first atomic bomb was created on Pueblo land. And the nuclear waste that resulted was buried in Pueblo sacred sites because US government agencies knew Pueblo people would never tell anyone because they won't say where their sacred sites are.

The reality is that Native nations have a longstanding intimacy with these kinds of economies, whether it's nuclear economies or fossil fuel economies. Understanding the historical conditions that force Native nations to participate in these economies is important, but I don't think it's a conversation about tradeoffs. It's about the fact that participating in these economies further entrenches us into the settler-colonial system—not just for our own dispossession, but also the dispossession of other people. The Red Deal presents an alternative: a shift away from the military-industrial complex and these extractive economies.

Does that entail a break from the Green New Deal that wants to see investment in green technologies that are also extractive? For example, you mentioned earlier that rare earth elements have to come from somewhere. How do technologies like lithium ion batteries that power electric vehicles fit into the Red Deal, given that lithium is mined on Indigenous land in Argentina, Bolivia, and Chile?

This idea that we're just going to continue the same level of consumption in a different economy is absurd because it requires the ongoing dispossession and subordination of not just Indigenous nations here, but also Third World countries. There are proponents of the Green New Deal that agree with this. We simply have to lower our levels of consumption. As Grace Lee Boggs and James Boggs said in *Revolution and Evolution in the Twentieth Century*, "The revolution to be made in the United States will be the first revolution in history to require the masses to make material sacrifices rather than to acquire more material things."

Many of the revolutions that erupted in the Third World *increased* consumption levels for the vast majority of those societies because they were under-consuming. To this day, there are many countries that are under-consuming. The United States is not one of those countries. Now, there are a lot of people in this country who are under-consuming, like Native people who live in dire poverty. But, by and large, the average North American middle-class and upper-middle-class person consumes way too much.

I don't dwell too much on settlers and whether they will ever have an ethical relationship to land. Some of them will turn into fascists — many already have — and some of them will follow us. If we're decentering whiteness, and we're decentering settler ontologies, and we're actually advocating for their abolition, what does that new world look like? What does ending the colonial relation look like?

Ultimately, we're trying to center what good relations to the land means. Instead of talking about car batteries, I think the real conversation should be: why are we working more than twenty hours per week? Why are there jobs that require air travel? Why don't we have a universal basic income across the globe so people don't have to leave their hometowns to find work? How do we end border imperialism so capital doesn't have an endless supply of cheap labor? Those are some of the things that I'm thinking about. ∿

What Green Costs

by Thea Riofrancos

Deep in the salt flats of Chile lies the extractive frontier of the renewable energy transition.

———

Clean energy advocates envision an electrified home running on 100 percent renewable energy with a Tesla parked in its garage, solar shingles gleaming on its rooftop, and a smart meter dutifully collecting usage data and uploading it to the cloud. But swim upstream and eventually you arrive at the extractive frontiers of the renewable energy transition.

It was 8:45 am on the first day of the 11th Lithium Supply & Markets Conference in the basement level of the W Hotel in Santiago, Chile. There was no way for me to blend in. "Providence College" on my name tag rendered me a curiosity. Still, I was glad I remembered to wear lipstick and that my backpack had straps that converted it into a tote.

I found an empty seat in the sea of suits, almost all men but of different ages. They hailed variously from China, Australia, Chile, the United States, the United Kingdom, and Argentina. They were market analysts and prospectors; equipment salesmen and regulators; executives, consultants, and peddlers of information in the notoriously opaque world of lithium, a "space," in Silicon Valley talk, not quite meriting the word "market."

As I slid into my seat, the chairman of one of the largest lithium companies in the world, with a sordid past in a corrupt privatization process under Augusto Pinochet's brutal dictatorship, took to the stage. "Mining is the spine of Chile; mining runs through our veins." I might have been the only person in the room who immediately thought of Eduardo Galeano's anti-colonial page-turner, *Open Veins of Latin America*—incidentally penned the same year Pinochet came to power, brutally crushing the dream of democratic socialism in Chile. But I don't think the chairman meant to call to mind the vampiric iconography of global capital. The dead sapping the living; the blood and sweat and tortured landscapes of extraction, especially in its colonial variant.

Arm Wrestling in the Atacama

Lithium is the third element in the periodic table. It is highly reactive and exists either in compounds with other minerals in rock formations, or in clay deposits, or dissolved as an ion in brine. It is also the active ingredient in the lightweight rechargeable batteries that power electric vehicles (EVs) and store energy on renewable grids. This is why lithium is essential for the coming energy transition.

In the United States, transportation is the single largest source of carbon pollution, accounting for about 30 percent of emissions. Achieving anything like a safe climate means we have to swap internal combustion engine vehicles for EVs, and hook up those cars, trucks, and buses to an electric grid powered by wind or sun. (Transitioning from a model of individual cars to one of mass transit would facilitate this process, and have other positive environmental effects.) Lithium enters into this equation twice. First, it is a material input for EV batteries. Second, batteries are in effect an energy storage technology, and grids that operate on intermittent gusts of wind and rays of sun need a mechanism to smooth supply and match it to demand. (Dramatically reducing our overall energy consumption would also help.)

> *"The dead sapping the living; the blood and sweat and tortured landscapes of extraction, especially in its colonial variant."*

The brines of Chile's Salar de Atacama lie 7,500 feet above sea level on an Andean plateau and supply roughly 30 percent of the world's lithium. These salty underground reservoirs are located beneath a closed basin ringed by the Andean mountain range. A perfect storm of climate, geology, and chemistry has concentrated lithium in the waters below the rugged surface of the vast Atacama salt flat, which in total measures about two-thirds of the surface area of my home state of Rhode Island.

But resource extraction throws this vulnerable desert wetland out of whack. Getting the lithium entails sucking up the brine at an astounding rate. SQM, the company whose chairman I heard at the conference, pumps out brine at a rate of 1,700 liters a second—95 percent of which is then evaporated. In other words, extracting lithium involves drawing up a lot of water and throwing most of it into the air.

Almost any corporate representative will say that extracting and evaporating brine has no effect on freshwater. But talk to any scientist or regulator familiar with the Atacama basin and they will tell you that the two types of water interact—that removing the brine eventually lowers the water table, threatening supplies needed for drinking and irrigation.

"Thirty kilometers away, swallowed by the horizon, stood the big lithium installations."

You can think about it like an arm-wrestling match. The brine water is underneath the salt flat. The freshwater systems are located at the flat's perimeter. The two kinds of water are separated by a dynamic interface: a surface tension generated by the fluid's differing density levels. Brine is much denser than freshwater, weighed down by dissolved elements like lithium. But while brine has the force of mass on its side, freshwater—which originates from snowmelt high up in the Andean peaks and the aquifers they feed—has the force of gravity in its favor. They are locked in a contest: mass versus gravity. When brine is removed, the interface separating them shifts towards the center of the salt flat, dragging freshwater with it—and away from the Indigenous communities located on the flat's perimeter.

Flamingos and Quince

I first saw the Salar de Atacama after driving around the mountains on the border of Bolivia. The Licancabur volcano loomed over us. We drove through a sandstorm, my first—memorable for its force, noise, and the way the suspended sand revealed the dance of the air's rapid movement—which was made stranger by being paired with a rainstorm. We drove through many microclimates. The vegetation completely changed with the elevation. Gaining in altitude, the cooler and wetter air supported denser life; scrubby patches gave way to lusher meadows.

Descending again, we entered the desert. Oases dotted the landscape: trees and shrubs congregated around the streams that flowed through mountain ravines. These *quebradas* are the basis of the built environment and social life of the eighteen Indigenous communities that neighbor the salt flat. Traveling through stone canals and filters, the quebradas feed small farms. The plots are enclosed with rustic wood fences and strategically planted trees for shade. They produce an incredible variety of produce. On a visit to the community of Toconao, I spotted figs, pomegranates, and quince along with the usual maize.

Driving further east, we reached the Los Flamencos National Reserve: an immense sweep of white and grey, rimmed by mountains in all directions. To our left was pure salt crust; to our right was the same crust interspersed with lagoons, where flamingos fed on tiny brine shrimp. The lagoons were tinted red in places, from the interaction of algae, sun, and wind. It was expansive in a way I associate with the ocean. The ground was crunchy and knobby beneath my hiking boots.

Just out of view were the zones of extraction. Thirty kilometers away, swallowed by the horizon, stood the big lithium installations. During the conference in Santiago, I had heard executives say that environmental protection measures should be improved—but also that there was nothing to be worried about. The rich ecosystem of these desert wetlands—the cotton-candy-pink Andean flamingos, white tufted grebes, and regal vicuña—did not figure prominently in the discussion. Indigenous communities were briefly invoked, and workers got a mention or two. But for most of the conference, the human and ecological texture of the *salar* receded from view.

Yet communities like Toconao are already feeling the effects of extraction on their everyday lives. Abnormally arid conditions reduce the streams' flow, constraining access to water for crops and drinking. And, due to global warming, the swings are getting more unpredictable: long dry spells are punctuated by mega-rains that destroy infrastructure and plants, and can't easily be absorbed by the soil. These changes also threaten habitats for wild vegetation and animals: biologists have found reduced species counts for the Andean flamingos.

For the suits at the W Hotel, the Atacama was an extraction site, an operational landscape, the beginning of a long trail of logistics and profit. But what of the vicuña and the quince, and the communities rooted in the flow of the desert's precious water? What would it look like to bring these into view?

"*Organizing across three national borders, in rural spaces crosscut by dirt roads and underserved by transit and WiFi, is hard.*"

Teeming, Crawling, Floating, and Flying

The day after my first visit to the salt flat, I met Ramón. We spoke for three hours, neglecting other appointments, over coffee and *medialunas de manjar de leche*, which are croissants filled with caramel made from condensed milk.

In contrast to many of the petit-bourgeois transplants in San Pedro, the mushrooming tourist hub in the Atacama, Ramón is from a working-class family on the rural outskirts of Santiago. He is a cofounder of the Plurinational Observatory of Andean Salt Flats (*el Observatorio Plurinacional De Salares Andinos*), a transnational network of environmentalists, concerned scientists, activist lawyers, and residents of affected Indigenous and *campesino* communities from across the Andean plateau known as the "lithium triangle." The lithium triangle encompasses parts of Argentina, Bolivia, and Chile, and contains more than half of the world's known lithium reserves — although members of the Observatory dislike the term for reducing their world to the resource extracted from it. (Full disclosure: I am a member.)

The Observatory pushes back against "green extractivism": the subordination of human rights and ecosystems to endless extraction in the name of "solving" climate change. Its platform affirms the broader cultural, natural, and scientific value of the salt flats — not just the economic value of its lithium.

It's challenging work. The Observatory is trying to weave together a novel organizational form, with ambitions at the transnational scale of extractive capital. But organizing across three national borders, in rural spaces crosscut by dirt roads and underserved by transit and WiFi, is hard. At the industry conference in Santiago, there were tensions between capitalists and the state, and between potential investors and mining firms. But on the whole, these elite alliances are relatively easy: lubricated by money and airplanes, smartphones and endless hors d'oeuvres. The obstacles facing international movement-building are much larger.

These obstacles were in evidence at an Observatory gathering at the University of Atacama in June 2019. The Argentinian delegation never made it: snow had rendered the border crossing impassable. The president of the association of eighteen Indigenous Atacameño communities, Sergio Cubillos, was likewise absent. The communities he represents, along with Indigenous groups throughout the country, were engaged in an all-out mobilization against Chilean President Sebastián Piñera, whose government was trying to further fragment and privatize Indigenous territory.

But those who made it to the gathering were able to help develop a different vision for the habitats and wetlands of the region—an alternative to the one being pursued by the suits in Santiago. This vision is vividly captured in the work of Portuguese artist Mafalda Paiva, which were displayed at the Observatory event. In her paintings, the salt flats hum with a preternatural vibrancy, an effect produced by the exaggerated density of species and radically foreshortened topography. This teeming, crawling, floating, and flying life was invisible at the Santiago conference but composed the emotional core of the Observatory gathering. Paiva offers a kind of eco-utopian hyperrealism—and orients us to a very different future than that imagined by lithium capitalists.

Common Futures

The Observatory opposes green extractivism because of the very real harm it inflicts on humans, animals, and ecosystems. But their position raises thorny questions about the renewable energy transition. As each dire climate science report makes clear, emissions from fossil fuels are rendering the planet increasingly unlivable. At the same time, building a low-carbon world carries its own environmental and social costs: every wind

Mafalda Paiva, "Salar de Atacama"

turbine, solar panel, and electric vehicle requires vast quantities of materials mined from the earth, transported in container ships over great distances, manufactured in factories likely still powered by coal, and transported again to consumers. This globally dispersed supply chain, like any other under capitalism, facilitates a race to the bottom, as capital perpetually seeks cheaper labor and cheaper nature.

Not all communities along this chain have a say in who bears the social and environmental costs, or how much effort should be expended to reduce them—unless they force the matter. The vaster and more complex the chain, the more challenging it is to mobilize across it. This global spread isn't new: the Industrial Revolution was enabled by raw materials extracted and harvested far from industrial centers. But in recent decades, technologies that disperse production even further have proliferated, from container ships to new trade agreements, computer-enabled "just-in-time" manufacturing to Special Economic Zones, making global capitalism an infinitely more intricate and interdependent web than Adam Smith ever dreamed.

When it comes to the renewable energy transition, how this web works has especially high stakes. It is a question of who controls our future. A world buzzing with hundreds of millions of Teslas (or worse, e-Escalades), made with materials rapaciously extracted without the consent of local communities, manufactured under a repressive labor regime in polluting factories—in other words, a world not unlike our own, but powered by wind and sun—is not an inevitability.

Other futures are also possible. The already unfolding energy transition offers a historic opportunity to dismantle the American lifestyle of privatized and segregated suburban affluence and build something better in its place. This lifestyle has always been a nightmare, ecologically and politically. The less energy we consume, the fewer raw materials we will need. This is not a call for eco-austerity: currently, energy consumption is highly unequal and wasteful. We can construct a society that is both low-carbon and plentiful in the ways that matter for most of us.

Doing so will require acknowledging how the material substrate of our lives is intimately, and often violently, connected to ecosystems and people beyond our borders. Trade, production, and consumption could, in theory, be reorganized to prioritize climate safety, socio-economic equality, Indigenous rights, and the integrity of habitats.

Yet achieving such an outcome will take political power, strategically deployed. Amid the overwhelming complexity of contemporary capitalism, it's easy to forget that supply chains are not the product of geographic destiny. Indeed, a key aspect of environmental injustice is that contaminating processes—mines, power plants, or factories—are sited where ecosystems and human lives are seen as disposable or deemed to lack political influence.

The corollary is that force from below can obstruct and even reshape global flows. This force is particularly effective when exercised at "chokepoints": points of obligatory passage for people and goods. In addition to the factory floor itself, the infrastructure of logistics (ports, ships, warehouses) and the sites of extraction (mines, rigs, refineries) are potential bottlenecks, and thus nodes of vulnerability for the system as a whole. In other words, they are strategic sites for disruption.

> ## "We can construct a society that is both low-carbon and plentiful in the ways that matter for most of us."

I might not know the exact shape of the world I want. The present weighs heavily and makes imagination difficult. But I know it starts with relating to this planet's bounty as mysterious, vital, and nourishing; envisioning abundance as shared flourishing; and broadening our solidarities to encompass people we may never meet and places we may never visit but whose futures are bound up with our own. The *salar* will thank us. ⁓⁓

Thea Riofrancos is an assistant professor of political science at Providence College. She is the author of *Resource Radicals: From Petro-Nationalism to Post-Extractivism in Ecuador* and the coauthor of *A Planet to Win: Why We Need a Green New Deal*.

In the Shadow of Big Blue

by Ellyn Gaydos

The birthplace of IBM is struggling to live in its shadow.

———

"It's kinda like the Parthenon now, it's a testament to something..."

—*Nick Bongiorno, former IBM temp worker and activist*

The IBM country club on a hill overlooking Endicott, New York has been empty for thirteen years. Now beyond repair, it was once abuzz with the activity of some 14,000 IBMers and their families. There were basketball games, swimming pools, a bar, a stage, banquet halls, guest rooms, and a golf course, all open to the thousands of IBM employees in Endicott. This was the town where, in 1911, International Business Machines was born.

IBM's Plant Number One manufactured punch-card tabulators in downtown Endicott. Then came typewriters, printers, and the System/360 computer, after which a parade of ever-newer models were made. For decades, IBM dominated the computer industry. It was not until 1996 that their market value was surpassed by Microsoft. They have since fallen far down the ladder:

these days, IBM is only the ninth-largest tech company in the world. They have gotten out of the messy business of making things and are now, primarily, a software and services company. In 2002, after more than ninety years, IBM ceased manufacturing in Endicott.

"It is hard for people to believe in invisible things, but the plume began to manifest. Its vapors started to travel underground."

Today, the IBM country club is utterly defaced. Defaced by the freeze and thaw of water, and the flooding of the Susquehanna River. Defaced by teenagers cycling through in the night, the keepers of no-man's land. This is where they come to get high and destroy the memory of IBM, their father's house. Graffiti on an unbroken window says *Did You Ever Love Me?* in searing pink paint.

> *Live*
> *Die*
> *Suck Me Softly,* it says.

In the old brick rooming house, white columns have fallen against the floor. The fireplace is stuffed with saplings and branches; it looks wild and evil. Stiff palatial curtains remain. There are pellets of rat shit on the ground and moss grows across the maroon carpeting. The pink and green wallpaper peels and blooms like wild roses. On the floor is a crumpled up brochure from 1954. "IBM Family Day" for "Plant Employees," it says, on a placard held up by a smiling clown. After pie-eating contests and a softball game, at 6 p.m. the IBM band would play.

In 2006, the Susquehanna River rose through the rooms, easily ripping at their seams and fixtures, a flowing push against the old giant. Named "Oyster River" by the Lenape people, the Susquehanna is one of the country's most polluted. It came to know nuclear meltdown from Three Mile Island, and it knew coal and fertilizer and feces, before it came into the club. Then the teenagers charged against the broken surfaces.

In the locker room there are still four inches of green soap in the dispenser but the porcelain toilets are smashed or filled with shit. The mirrors have upside-down crosses and

666
PUSSY
SATAN

With language learned from horror movies, the trespassers hack away at Big Blue's shadow. A fuck you, from the town to IBM. A mutual fuck you, from IBM to the town.

The Plume

It started beneath a swath of train tracks and poured gravel beside Building 18, where circuit boards were made. It formed from repeated spills of volatile organic compounds used in the degreasing and cleaning of microchips. In 1978, 1.75 million gallons of wastewater were released. That same year, 4,100 gallons of liquid solvents, including trichloroethylene (TCE) and trichloroacetic acid (TCA), were released. In 1980, IBM contacted the Environmental Protection Agency (EPA) to report that contaminants had begun to form a pool beneath Building 18. This is what became known as the plume.

It is hard for people to believe in invisible things, but the plume began to manifest. Its vapors started to travel underground. It spread to encompass 300 acres of the town: churches, movie theaters, grocery stores, and 480 homes. It was no longer invisible when people began to get sick.

The danger of TCE exposure is that it is carcinogenic and can impair fetal development. The chemical penetrated deep into the groundwater as a liquid and then began to evaporate, moving through air pockets in the soil. This migration continued through cracks in the foundations of homes and buildings, creating an indoor environment of prolonged exposure. People who both lived and worked in the plume were called "double dippers."

When the initial spill occurred, IBM began digging wells. Twenty extraction wells pumped out contaminated groundwater. In 2002, the year IBM shuttered its factories, the New York State Department of Environmental Conservation (DEC) required the company to test the air quality. By 2004, they entered into a "formal consent order" to investigate and remediate the contamination. What IBM found led them to install vapor mitigation systems in homes and other buildings in the plume.

These systems are discernible by the white boxes attached to long pipes that reach up to roofs, rerouting the vapor from underground into the surrounding air outside. Now in homes, houses of worship, billiard bars, and barber shops, there is a constant whir of ninety-watt motors working against TCE. The contamination is continually announcing itself. It is ignorable as a low drone, forgotten and re-heard over and over again.

Endicott is important because it is not unique. It is a story that one can almost write without knowing the specifics. It is a story of the postindustrial long after the last shoe, car, or computer travels through the factory.

Endicott proves that it is only through extraction, refinement, and manufacturing that computational feats of any kind are possible. The machine is made of materials from the earth: copper, gold, nickel, silicon. In order to purify, clean, and combine its pieces, intensive chemical baths are used. The computers and smartphones that result have an incredibly short working life, on average just two to three years. A shorter life than a car tire, a winter coat, a stereo, or a shovel.

Though compact when presented to consumers, these devices also have a huge material footprint. The inputs for a microchip are 630 times the mass of the final product. After the product is made, all of these excess inputs recombine into new chemical slurries, the unsaleable byproduct of the machine. These life-altering chemicals return to the earth in indigestible ways, and creep through our basements, waterways, genomes. There are 2.71 billion smartphone users in the world, and 1.5 billion personal computers in use. This means there are many towns like Endicott.

"*Endicott is important because it is not unique.*"

Inside the Clean Room

I too am from an IBM town, one in northern Vermont, the only in the state. Like thousands of other Vermonters, I worked in the factory there. I didn't get sick and neither did my immediate coworkers, but I began to hear troubling stories. I also began to read article after article imploring IBM to stay in Vermont. Eventually IBM did leave, but unlike in Endicott, the factory was taken over by a new company and kept running. In Vermont the pollution was quieter. The factory was not classified as a

Superfund site, so it did not stick in the public conscience—only in the thoughts of those who worked there. The pollution also remained quieter because the factory is still in operation.

The work itself, twelve-hour shifts in a factory built to protect the product not the people, was dehumanizing. I performed one step in hundreds required to make microchips. The Vermont plant specializes in amplification chips that transmit signals to satellites and enhance the speakers nestled next to our ears on our phones. Twenty-four hours a day, white-clad employees walk up and down the fluorescent hallways of the factory: workers in hoods, gloves, veils, booties, and coveralls so that the eyes are all that is visible. This is to protect the delicate chip from human contaminate.

> "People had sex behind the robots, got loaded on their lunch breaks, defecated in trash cans, and hid six-packs in the floor tiles."

I worked in the "wets" department, applying chemical cleaners to microchips so that layers of circuitry can be built cleanly on top of one another. The chemical wash machines I operated look similar to a home washing machine. The workers used to man the chemical baths themselves, balancing the boxes of wafers on a hooked pole and removing them when a manual timer dinged. Some even used to volunteer for this job because they thought it could get you high. Now the process is more mechanized: I performed the simplified and repetitive function of loading and unloading the chemical wash machine, then putting the product onto an automated overhead track, so that there were always half-made microchips floating above, traveling to the next tool required to complete them.

The pay is a few dollars above minimum wage. I had to pass background and literacy tests and formally agree to chemical exposure in my job interview. I'm not sure whether it was the low pay, volatile hiring and firing, or the sensory deprivation inherent in working in a "clean room," but the workers engaged in small rebellions. People had sex behind the robots, got loaded on their lunch breaks, defecated in trash cans, and hid six-packs in the floor tiles. One janitor, tasked with cleaning the factory, called in a bomb threat.

The Vermont factory, like the one in Endicott, spilled chemicals into the surrounding area. Five miles of underground piping leaked, delivery drivers spilled solvents, workers poured waste into drains that were not hooked up to pipes, the contents of unsealed wells and a "sludge landfill" seeped into the earth. The chemicals spread forty acres wide and travelled 300 acres deep into the bedrock of the town.

In 1979, the contamination came to light when IBM reported the spills to the EPA. TCE and PCE, another carcinogenic degreasing agent, were found in the nearby river and lake: in some areas, PCE levels were 19,000 times higher than the state standard allowed in drinking water. In cooperation with the EPA, IBM began a groundwater cleanup campaign. Contractors for the company in Vermont said the process, conducted across six sites, would take more than two hundred years. Scientists say it could take thousands.

In Endicott, the New York State DEC does not have a time estimate at all. In both cases, the contamination will cling to the land long past the lifespan of a factory, a product, or even an industry.

Sacrifice Zone

When IBM left Endicott there was no new company to take over the factory, like in Vermont. In the newly quiet town, IBM's legacy began to ring louder. It was not simply the land that was changed, but the people too.

In Endicott, both teenagers and adults got cancer. One girl broke her leg walking through the halls of her high school. At the hospital she found out she had bone cancer. And there were children born with malformed hearts. The New York State Department of Health (DOH) reported fifteen cases of infants born with heart defects over seventeen years in one Endicott neighborhood. This number is more than twice that of the normal population.

When Kevin Every moved to a rental in Endicott from Philadelphia, his wife Tiah was pregnant with their youngest. If he had known that the house was in the plume he never would have rented it. "But nobody gave me that choice," he said. Instead, he found out about the plume on the news. When his son Deron was born, he had six different heart defects. At thirteen days old, he had his first operation. At eight years old, he had a stroke.

"He prays," Kevin told me. Deron prayed for a new heart, and got a transplant this summer. The family eventually bought a house outside the plume. Kevin doesn't know if ventilation was ever installed on his rental or who lives there now. Renters, who don't know the history of the area or can't afford higher rents, are unfairly affected. They are often transient enough not to be accounted for when they get sick. At the Ronald McDonald House in Syracuse, when Deron was an infant recovering from his initial heart surgery, Kevin met seven families. They were his neighbors from Endicott. Their babies also had heart defects.

When Kevin got back to Endicott, he called a lawyer.

Today, Kevin spends a lot of time traveling to doctors. He tells his four other kids, "If that was you, I'd be doing it for you." I asked Kevin about IBM: if he thinks it's an accident, if he thinks they're sorry. "They can do whatever they want just so they can have a buck… families lose because now they have a loved one that's sick. If I went out and changed the oil in my car and dumped it on the grass I would get in trouble," he said. When a commercial for IBM comes on TV, he can't bear to watch it. "This is happening all around America," he said. No matter if they ever admit what they've done, "right is right and wrong is wrong."

James Little worked at IBM in Endicott for fifteen years as a senior operator making chip boards. He worried about leaky machines. Once, he shut down a machine that was spilling chemicals into an overflow tray. His manager chastised him and told him never to do that again. James heard rumors of chemicals dumped in holes in the concrete cellar, pipes with leaks, and train cars that spilled their deliveries of chemicals. Workers around him were getting sick. A girl who worked beside him got a brain tumor. A man in his department had his nostrils "eaten out" from the fumes. "The bottom line," Little said "was they wanted to get the work out… I think people were sacrificed." Little became an activist and workplace safety advocate. He talked to the press. His manager told him if his name appeared in the paper one more time he would be fired. He kept his job until the factory shut down.

Such stories aren't limited to Endicott. There are similar stories wherever IBM manufactured chips. Michael Ruffing and Faye Carlton worked at IBM in East Fishkill, New York. They sued IBM after their son was born blind with facial deformities that prevented him from breathing normally. Candace Curtis, whose mother worked while pregnant with her in the same East Fishkill plant, was born without kneecaps. She is not physically capable of talking. Nancy LaCroix, of IBM Vermont, had a baby girl with bone defects, which caused her brain to protrude from her skull and left her with stunted fingers and no substantial toes. One unnamed child of an employee was born without a vagina.

Superfund Site, IBM on Trial

Beginning in the 1990s, lawsuits began popping up all over the country where IBM made chips. A case from IBM San Jose that sought to establish a cancer link with chemical exposure in factories was dismissed when, after two days, the jury decided in favor of IBM. More than 200 workers in Vermont, New York, Minnesota, and California brought lawsuits against IBM for work and resultant environmental conditions that caused them or their children to become ill. All settled out of court.

In 2008, a group of around 1,000 Endicott residents sued IBM for $100 million over increased occurrence of kidney cancer, heart defects in children, and lowered property values. In proceedings, IBM was forced to disclose the contents of their "Corporate Mortality File," a database dating back to 1969, a decade after the invention of the microchip. IBM claimed the file was created to track pensions and other lasting benefits to the families of deceased workers. It contained 33,730 former employees with basic identifiers like sex, age, work history, and, most importantly, cause of death. Increased rates of respiratory and breast cancer as well as cancer in the internal organs were found.

After seven years and no trial, IBM eventually settled the case out of court for an estimated $13 million. No wrongdoing was publicly admitted and no cancer link credibly established.

"These are tragic cases, but there is no scientific evidence that there are increased rates of diseases of any kind among IBM employees," an IBM spokesman stated in response to the rash of lawsuits. Kevin Every's family was part of the $13 million settlement. He can't go into detail but said, "We didn't get what we shoulda got. They asked me how much I think we should get, I said everything. [Deron] can't even go on the playground."

> "*Candace Curtis, whose mother worked while pregnant with her in the same East Fishkill plant, was born without kneecaps. She is not physically capable of talking.*"

Although IBM has denied responsibility for the health problems in Endicott, they have committed to helping clean up the town. (They have also tried to burnish their public image and defuse anger among residents with philanthropy: in 2002, on the day Endicott was classified as a class 2 Superfund site by the EPA, IBM gave Endicott a "gift" of $2 million.) The plume has shrunk considerably since remediation efforts began pumping out contaminated groundwater. A smaller plume means less toxic vapor intrusion into local homes and businesses. For now, the white vapor mitigation boxes on the outside of houses remain and the groundwater pumps continue to suck up polluted water.

Fortunately, IBM is a rich company with plenty of money for remediation. For IBM, spending $270 million on environmental clean-up projects across the nation in 2017 was easily absorbed in the following year's revenue of $80 billion.

James Little, the former factory worker, hopes that new business will come to Endicott now that it is cleaner. He still loves his hometown. "I would consider this site pretty much safe," Little told me, but he knows the spilled chemicals will be nearly impossible to totally eradicate. They will continue to use the vapor mitigation systems and the groundwater pumps.

So much polluted groundwater was pumped out (over 800,000 gallons), in fact, that sinkholes began to form in the dry soil. It is like rinsing and squeezing a sponge, Little told me. The same flood that partially destroyed the old IBM country club helped to flush out some of the contaminants from the ground. Still, the chemicals bind to the dirt, and it seems unlikely that they will ever be totally eliminated.

Beyond the borders of Endicott, there remain seventy-six microchip manufacturing facilities in the United States. There are many more around the world, from South Korea to Taiwan, Germany to Singapore. Toxic TCE is not just a problem for IBM neighborhoods, then, but for computer manufacturers all over. Of the Superfund site National Priorities list, TCE is in 1,045 of 1,699.

Around the same time that IBM pollution came to light in the United States, manufacturing was being shipped overseas, along with the pollution. Even as dangerous chemicals like TCE are replaced, or in rare cases outlawed, the sheer demand for the product is perhaps the biggest danger. Production is valued over safety and product is prized over resources. Until this equation changes, Endicott will have many sister cities.

Life Goes On

It is 7 a.m. in Endicott, the time the morning shift at IBM would start if the factory was still running. The light outside is a familiar arctic blue which makes the trees and snow seem both flat and harsh. I'm hungover, sitting on a bed in the Endwell Motel.

I spent last night at a bar called Close Quarters outside the reaches of the plume. My friend Sarah came here with me to take photographs, and she came to the bar too. Sarah drank mini bottles of white wine while I drank Labatt Blues. We shared a basket of tater tots, talked, and watched football on TV.

> *"Around the same time that IBM pollution came to light in the United States, manufacturing was being shipped overseas, along with the pollution."*

We met an Italian guy, "Tim," from here (townies call it "End-y-cott") with a triangular nose and a tattoo of a big cumulous cloud spilling over his right hand. He bought me a shot of Jameson and I told him about my research. He said he didn't work at IBM but he knew people who did, real old timers. He's too young to have worked at this IBM. Instead, he drives an 18-wheeler for CVS.

Tim used to be a drug addict like a lot of people I meet around here. They are frank about it: sick but better now. It's not so much opioids that are the problem these days but meth. It seems the desire is not in wanting to slow down, but instead to speed things up. The town population has dipped to 13,000, smaller than the IBM workforce at its height. Instead of manufacturing, people now work in retail, healthcare, or the service industry. Nearly 20 percent of the town lives in poverty.

Tim showed me videos of his fish tank on his phone. I kept asking, "Is that a goldfish?"

"No," he'd say. They all had different species names, bright yellows and oranges swimming across the slick expanse of his phone. He showed me his truck delivery route for the next day. He'd be driving north to Vermont, right by the factory where I used to work.

We began to talk of violence. He likes to fight, he said, smiling, "but when I see a flower, I see a woman." That's why men *need* women, he told me and Sarah. We traded stories about killing pigs. Tim's story was about a butcher crying and shooting a pig that wouldn't die. He cried so much he was blinded by his tears. He cried so much he started praying. Sarah and I told him about the pig we shot that ran away into the woods, how it had to be tackled and shot again. The instinct in nature—a flower, a pig, a town—that does not want to die was there both times. We talk like this about life and death as a way to talk about the poison of the plume, and our hope for the future. ⌇

Ellyn Gaydos is a farmer in New York and is writing a book of stories on the nature of seasonal work.

Lab Cultures

An Interview with an Anonymous Biologist

In the last decade or so, advances in gene-editing techniques have allowed scientists to modify genetic code with an unprecedented level of precision and speed, and at a fraction of the cost. This has been game-changing for biological research. Scientists can now make and test changes to genetic code in ways never before possible, with far-reaching ramifications.

The scientific discoveries that get buzz in the popular press and what scientists themselves are excited about are often pretty different, however. So we were curious to learn more about how scientists see these developments. Even more fundamentally, we wanted to know about the nuts and bolts of research: how do labs operate? Who works in them, what is the hierarchy like, and where does the money come from? How do conversations about ethics in scientific research differ from those in the tech world? What job prospects outside of academia are out there for someone with a PhD in biology?

Over drinks at our kitchen table, we sat down with a current PhD student in cell biology to learn more.

You previously worked as a software engineer for a large company in Silicon Valley. How did you decide to become a PhD student in biology?

I remember finally reaching that decision while I was reading this Richard Dawkins book, *The Ancestor's Tale*, which is an exploration of the tree of life. It works from humans backwards, and tries to find our common ancestors with all the other organisms that exist. I had been enjoying the book and all the little vignettes about life, ignoring his occasional rants about why God is dead and all that. Eventually, he got to bacteria, and I started getting super amped about all the things that he was saying about bacteria and viruses and archaea and whatnot.

 It got to the point where I would strike up conversations with random people on the bus about what I was reading, and it dawned on me that I was excited about biology in a way that was distinct from the way that I was excited about my job at the time. I had wound up in a discipline where I got to solve cool problems, but I wasn't really engaged with the physical world in any meaningful way.

To sanity-check my desire to change careers, I did informational interviews with biologists to get a sense of the possible jobs that would let me touch this stuff in my day-to-day work, and what I would need to do in order to have my hand on the tiller.

One of the things that quickly became clear about the field of biology was that people typically had answers to *basic* questions, but that there were a ton of questions no one had answers to. When I'd ask a deeper follow-up question, they *might* have an answer. And then, when I'd go one level deeper, they'd be like, "Oh yeah, nobody knows that." Eventually, it became clear that I really needed to have a PhD in order to have the conversations that I most wanted to have, so I ended up pursuing that.

"The idea of tons of tiny agents interacting at a scale where the physics that we're used to don't even apply — that's a super exciting world to think about."

It's funny: when I was on something like the ninth iteration of my application essay, I was sitting on the phone with my mother in a cafe in the Mission trying to justify this transition from computer science and artificial intelligence into biology, and having a hard time coming up with a cohesive connection. But she was like, "This makes perfect sense. When you were ten, you were excited about nanotechnology and talked about how you were going to have tiny robots clean your teeth, and now you're talking about bacteria in the same way."

If I had to trace the common thread, it would be the idea of tons of tiny agents interacting at a scale where the physics that we're used to don't even apply — that's a super exciting world to think about.

So now you're a PhD student working in a research lab. What does your work there look like, generally speaking?

We study bacteria, so we wind up learning stuff that's also relevant to people who want to prevent infectious diseases. That said, we're not a pathogen lab; *we're* not focused on preventing infectious diseases. We certainly collaborate with labs that are — the ones that are working on pharmaceuticals and antibiotics, or that study pathogens and try to understand their biology. We can say things like, "We've now identified collections of genes in bacteria that, when you remove them, kill the bacteria," and then somebody more directly working with pathogens can say, "Okay, we know that we've got pharmaceuticals that remove this one gene function in bacteria, others that knock down this different gene function in bacteria, and they've both been approved by the FDA."

When publishing research papers, my lab is always very interested in what we call "having biology in the paper." I'm sure that sounds tautological and opaque if you're outside the discipline, but it means if I build a new tool, I want to be able to show you something new about the way that the biological world works; I don't want to just describe the new tool I built. We're often trying to develop new technologies that let us make robust and refined measurements about bacteria, and then elucidate the function of the elaborate web of genes driving life at this micro scale. When we can do that, that's a great outcome.

> "*For us, mostly we just care about how bacteria do bacteria, because it's fundamentally fascinating that these tiny things are able to survive the vicissitudes of the world around them.*"

It's hard to understand what bacteria are doing and how they're doing it just by looking at them. You can see these little pill-shaped things, and if we mess with them we can sometimes see them die in kind of interesting ways. But at the level at which we can watch bacteria, trying to make conclusions about them would be like trying to evaluate the well-being of France by counting the number of national monuments that were built over time in Paris—at this level you have no visibility into the lives of individual people or what the main commodities are or things like that.

And so rather than directly observing bacteria, a lot of what we're doing is trying to create ways to perturb them, and then measure and record their lives in large datasets of observations. We want this data to be sufficiently nuanced that we can get a richer understanding of what is going on when we analyze it later. That then provides tools for people to study a wide range of bacteria—pathogenic or otherwise. People use bacteria for all kinds of purposes, so for example many of the techniques we develop can be used to figure out better ways to get bacteria to produce stuff, whether that's food products like alcohol and yogurt, energy products like methanol and ethanol, or enzymes like the ones in your cleaning detergents.

But for us, mostly we just care about how bacteria do bacteria, because it's fundamentally fascinating that these tiny things are able to survive the vicissitudes of the world around them, where a slight shift in the salt concentration in water is this massive life or death situation that happens over the span of, like, three seconds. It'd be like your entire neighborhood being instantly flooded. Bacteria just have a program to deal with that. They don't have a computer, they don't have a brain—they're smaller than one brain cell or a single transistor, and yet they've some- how got a plan of action to deal with that scenario in real time.

I find that fascinating, and I think that's what brings a lot of people into the kinds of disciplines where I reside. And the same thing is true of the individual cells in the human body, and a lot of people who are "working on a cure for cancer" are mostly just fascinated by how in the hell these elaborate tools called proteins are being assembled by the endoplasmic reticulum and the Golgi bodies and whatever. There is this enormously complicated thing that's happening a gajillion times per second in your body.

The way you get money from the National Institutes of Health (NIH) is by saying "and if we figure it out we'll be closer to curing cancer"—and don't get me wrong, the fraction of people who actually care about that in the biology community, including my lab, is non-trivial. A lot of people went into biology to like, I don't know, avenge their grandfather or something. But for me, in the war between bacteria and humans, I'm probably gonna side with the bacteria. I'm not really in it for human health. It's great if I can find funding by helping to facilitate that, but I'm mostly just curious how bacteria are doing their thing.

In terms of the research you're doing now, what is the best-case outcome of what you're working on?

Graduation. *(laughter)*

My lab is very focused on basic science, which is to say that the goal is explicitly open-ended—it's not like we either find the answer to a specific question or bust. We have a pretty good idea of where there are interesting answers to important questions, and so we come up with new ways to shine light into those dark corners, and then hope we find cool stuff. But we aren't too dead-set on what that stuff is. Indeed, the story that we're telling about my PhD work has changed pretty dramatically just in the last year, and I don't think that's atypical.

The way the course of research in a lab pivots is a bit different from how a startup might pivot, however. If a startup pivots, they are changing their plan from one thing to another. Whereas in research, it's more like realizing you could do something that wasn't possible before, or realizing that a question we had abandoned in the past turns out to be relevant to a number of other things. When those discoveries happen, you refocus in a new direction and see where it takes you. You need to be agile to find the best outcomes.

So I think the "best case" depends a lot on where the light winds up pointing us.

> ## "In the war between bacteria and humans, I'm probably gonna side with the bacteria."

You Can't Come in, Clive

The main biological thing I've been hearing about recently is the advances in our ability to edit genetic code using CRISPR, which has a lot of hype surrounding it. Is there hype among biologists about this in the same way? Does it feel like a real thing with potential, or is that far off?

CRISPR is actually just an acronym for "Clustered Regularly Interspersed Short Palindromic Repeats." What the acronym means in and of itself isn't exciting; it's basically just a placeholder for a certain pattern we see in DNA.

We've sequenced DNA from lots of people, but even more bacteria and viruses. We don't know what most of it does, but identifying genes is way easier than figuring out what they're doing. It's fairly easy to identify the boundaries of genes themselves, and we're generally able to identify the basic structure of an organism's genome, especially in prokaryotes like bacteria that don't have more complicated genetic structures like introns. I mean, there's still a fair amount of complexity in bacterial genomes with things like intergenic DNA and transcription regulators, but it's nowhere near as complicated as eukaryotic genes in organisms like humans. Almost everything in a bacteria's DNA is devoted to something that will actually be turned into a protein.

But within bacterial genomes, scientists would find these patterns where it wasn't clear what they were doing. This pattern was given the CRISPR acronym, which basically says, "We see this banding pattern in the DNA where there's a repeated palindromic sequence, and then it has a bunch of other stuff in between the repeated palindromic chunks, but that's all we know right now." Over time, people put together that the stuff in between the palindromic chunks seemed to match DNA that had been found in viruses—kind of like a virus's fingerprint, wrapped up in the bacteria's DNA.

How did they figure that out?

The breakthrough for understanding this came from yogurt manufacturing. Danisco is the company that makes Dannon Yogurt, and a big part of their business is to culture bacteria in giant vats in order to make their yogurt. But their bacteria can get sick: there are viruses called phage that infect them. If a phage infects a bacterial strain that's in one giant monocultural vat, all the bacteria in that vat are getting sick. They're all fucked, and you just have to nuke the whole thing. All the money you've put into that batch is just gone.

So Danisco was like: how can we better identify when this is going to happen, and reduce the frequency of these infections? Because we would like to *not* lose millions of dollars worth of dairy product whenever our vats get infected. They had a research division that was studying the yogurt bacteria's CRISPR pattern to try to better understand how and when bacteria get sick. They saw viral DNA in this particular CRISPR pattern and decided to dig deeper.

"The CRISPR pattern is basically like a wall of Polaroids."

It's been a long time since I looked at the original papers, but they basically did experiments where they showed conclusively that CRISPR was involved with an antiviral protective mechanism. When the researchers infected the bacteria with viruses, the ones that successfully fought off the virus were modified genetically in such a way that their CRISPR pattern then contained clear genetic references to the virus that they just fought off. Furthermore, bacteria without the protective mechanism were vastly easier to infect with viruses. What they found was not just a general protective mechanism—it was a reactive, adaptive immune system in bacteria.

How does that mechanism work, exactly? How does the bacteria fight off the virus?

Let's say you go to your favorite nightclub. The bouncer in the back room has a wall of Polaroids with people's faces on them. When he gets out front and sees that jackass Clive trying to come back into the club again, he remembers Clive's face from the wall of Polaroids and keeps him out.

The CRISPR pattern is basically like a wall of Polaroids. It turns out that there are also all these genes that are adjacent to the pattern that are called "CRISPR-associated" or "Cas" proteins that vary in their details. There are actually lots of different CRISPR systems and patterns with different associated proteins; it's an immensely elaborate network that is used by bacteria against viruses, by viruses against bacteria, by bacteria against themselves. It's an incredibly complicated ecosystem, but fundamentally the idea is this: you have a bacteria's immune system standing there acting as a bouncer to suppress the production of some gene, either in the same organism or a different organism, and they're doing it based on this wall of Polaroids that they scanned when they started their shift. When a new viral infection happens, a lot of bacteria will just die. But the ones that manage to survive get a snapshot of the thing that almost killed them, add it to the wall, and the next time around when that guy shows up, they can be like, "No Clive, you can't come in."

So what practical applications does this mechanism have?

People realized early on that they could apply this in many different ways. It's very common in molecular biology, particularly for people who study viruses and bacteria, to make connections between learning something new about what bacteria are doing and developing a new tool that we can apply to other areas of biology. For example, useful biochemical research techniques like making use of restriction enzymes and polymerase chain reaction (PCR) and ligases... all these technologies ultimately come from various viruses or transposons or bacteria, where we find a thing that's doing something unusual like surviving in a circumstance we wouldn't expect them to. We can trace that to some weird molecular thing that's happening that we can then replicate.

When it comes to CRISPR, some researchers like Jennifer Doudna at UC Berkeley understood the import early on—there's a lot of dispute over who got there first, and who should get the Nobel or the patent or whatever—because the researchers saw, hey, we could use CRISPR in combination with a protein in the system that acts like a pair of scissors, which would allow us to use the combined system to start cutting DNA in extremely specific places.

How would that work?

Going back to the bouncer analogy, if I want to change the behavior of the system, I can add a new Polaroid to the wall without having to kill the bouncer or replace him. I just have to sneak in and add the new Polaroid, and now he's not going to let Jenny into the nightclub, even though Jenny has never done anything wrong. I'm just interested in seeing what happens when she's no longer allowed into the club. In the same way, with CRISPR gene editing, you can make targeted changes to extremely specific places in DNA with very little effort.

This has greatly reduced the amount of money and effort that goes into modifying genes or knocking down the genetic expression of a virus or protein or something. We can now make edits that target arbitrary proteins, and proteins are responsible for most of the physical action inside a cell. So that's really broadly exciting to biologists, and it's transforming *everything* at the bench.

Where it gets public buzz is that it's also potentially exciting in terms of biomedical applications. I think the average person wouldn't be excited about most of the things the biology researchers are excited about. There's huge buzz in the news, and there's huge buzz in the labs, but they are about different things. They overlap because there are researchers who do care about the fact that you might eventually be able to use CRISPR to cut genes that would disable cancer without killing the person in which the cancer exists. Or you know, God help you, try to make babies smarter.

The thing that I find most exciting and most plausible in the relatively short term is fixing coherently understood genetic diseases—being able to fix a genetic disease where I know that my partner and I both have the same predisposition for a genetic disease, but if we had a functional copy of this one gene we'd be fine. And so if we can modify embryos in vitro using CRISPR gene-editing mechanisms, that's a potentially exciting way to cure diseases before they start. There are people I care about for whom this would be a huge deal.

> ## "It doesn't just make science easier; it makes it profoundly easier, such that whole new categories of things are now possible."

Some diseases you can cure by having a finite number of cells that do the right thing, even if they're surrounded by cells that are doing the wrong thing. For example, if you've lost the ability to produce white blood cells due to any number of diseases, and I'm able to fix some of your bone marrow, or if I can give you insulin-producing genes inside your pancreas that is otherwise totally dysfunctional, then I might be able to make your life much better.

I think the media hype comes from this idea that we can probably now make modifications to an individual cell before it becomes a human being, and soon—or maybe already—we can modify cells or push pre-modified cells into an existing human being. That has the potential to fix problems that were previously unaddressable.

But that's still really, really hard to do, and it's not very illuminating relative to the real scope of what CRISPR can do. In basic science applications like the ones that I'm developing as part of my PhD, the excitement is more around using this system to engender outcomes that weren't previously possible in the lab. It doesn't just make science easier; it makes it profoundly easier, such that whole new categories of things are now possible.

Playing for the Lakers

Working for large software companies and working in academia have very different reputations around working conditions and generally how work gets done. Did you experience some culture shock in that transition?

Software companies and academia are different in a lot of ways. My particular program is kind of a weird home-for-lost-toys kind of a program, insofar as they were actually trying to recruit people who were not biologists by training. So I didn't feel as on the outside by virtue of not having a biology background as I expected to.

In terms of the actual disciplines being different, there is a big difference between engineering and science in terms of the way that people talk about problems. That's been the hardest bridge to cross.

One thing that I've noticed is that when confronted with a problem, I'll try to bound the problem. My thought process is: okay, there's this problem space we are trying to work in, so let's figure out the best-case and the worst-case boundaries, and then shave away at them based on facts that we know until we get down to the space in which things could actually exist. Then we can do experiments or develop algorithms to refine that in the direction we want to go—knowing that we're working towards this boundary where we can't do any better, or this boundary where things can't get any worse.

That feels to me like a very reasonable way to go about things. But, working within a scientific discipline, I find that when I start to frame the worst-case scenario, people will say, "What are you talking about? Things are *not* that bad. You have to calm down." And I'll start framing the best case scenario and people are like, "Well that's insane, and it's a cute thought, but things are *never* going to go that well." I'm still trying to figure out how to communicate in words that don't send people into a panic thinking that I am totally untethered from reality.

"Being a grad student is a fundamentally shitty situation to be in, both financially and existentially."

That's been the most jarring shift—not necessarily the biggest shift, but the most jarring shift. Some shifts were expected. A lot of the work that I do is still data analysis, partly because of my background and what I gravitate towards, and partly because that's the direction the field's going. There's still a lot of time sitting in front of a computer: coding scripts to plan complicated protocols, writing a paper, generating figures or other visualizations, analyzing datasets.

But in biology there's also a significant fraction of your time spent standing up in front of a flat surface with a bunch of—you know, it looks like what you imagine science looks like. You've got beakers of clear liquid that all look the same, and probably smell the same, and have totally different stuff in them, and you're trying to keep track of them and put them through very different chemical processes, and track a bunch of different readouts, and use ultra-precise measuring tools. It's like super-high-precision baking or something. So, as you can imagine, someone who works at Tartine has a different experience from someone who works at Facebook. And if you're in a biology lab, you're moving back and forth between those things, often in the same day.

So that was hugely different and I saw that coming, but it was still *terrifying*. When I was working in a lab over the summer before my program started, I turned to the guy I was working with and I was like, "Here's the deal: I do not know how to do any of this. I would love to learn about the science you're doing and contribute meaningfully, but mostly I need you to train me to do the most basic shit that you learned to do in Biology 101."

On top of the way people work together, and the basics of the work itself, I imagine there was a big financial transition as well.

Being a grad student is a fundamentally shitty situation to be in, both financially and existentially. If you are in a PhD program you are almost by definition working on things that nobody else cares about—except, if you're lucky, your advisor. But let's say you're working on something that lots of people do care about. In the best case, you're in a high-pressure race to be the first one to do it and you're terrified every day that someone else is going to scoop you. Meaning your work becomes irrelevant because someone else got to it before you — thanks, try again.

And that's horrible. I think the fact that you're barely getting paid anything is pretty minor compared to that. We live in the Bay Area, which is awful no matter what you do — you have to be rich to be poor in this town. Living on a grad student income has been painful for me, and I had a bunch of savings because I was extremely fortunate to be in the tech sector during part of the boom, enough savings that I could buffer my passage through grad school. Even so, I'm running out of money now, so it would be good to get back to a real salary — I don't know how the hell these kids who I've been going to school with who came straight out of college do it. It's a difficult existence.

"For many people, the main takeaway from doing a biology PhD is just that you got to be involved in science for some period of time; you do some really cool research and then you go do something else totally unrelated to biology."

What are your job prospects once you come out of a biology program?

Here's a roundabout answer to that:

The sense I get is that sometime in the 1980s or 1990s, the federal government put a whole bunch of money into biology.

Someone decided that we needed more expertise in biology to come to the United States, and the way to do that was to double the funding to the NIH—and they did that overnight. The result was this surge of universities creating aspirational biology programs, quickly building buildings, and hiring people to fill previously nonexistent departments.

Initially, that was a huge shot in the arm for the discipline. But at some point, there weren't enough places to install new faculty, even if you had people who were competent enough to fill the positions. We ran out of the NIH having extra money around to create new biology institutions, so we ended up with fewer biology faculty positions than postdocs, and fewer postdoc positions than graduate students, and so forth.

There's basically an infinite amount of good biology to be done, but the discipline in its current form can only support so many biologists. There are something like five times as many people who graduate with PhDs in biology per year as there are total faculty positions, much less positions that are open right now. So becoming faculty at the kind of institution where you're likely to get much NIH funding and be able to do cool research and live in an area that you want to live in, even after having studied at one of those top-tier research institutions, is like being in middle school and deciding you're going to play basketball for the Lakers. You're going to have to be really, really good to play for the Lakers, whereas most other people in your position will end up playing basketball in high school and that's the end of the story.

The aspirational thing on the academic side is to do a post-doc, where you continue to be an underpaid, overeducated researcher doing the actual job of research for another two to five years. That's not an entirely raw deal because it comes with a lot of freedom. But beyond that, academia is a really tough row to hoe. You're still trying to play for the Lakers. I'm at a sufficiently high-profile institution that there are people in my lab for whom it's not unrealistic to think that they might actually become faculty. It's not what I want for myself, or something I think I would be likely to achieve if I did want it, but it is something that people in my lab are actively pursuing.

For many people, the main takeaway from doing a biology PhD is just that you got to be involved in science for some period of time; you do some really cool research and then you go do something else totally unrelated to biology.

The other option that allows you to stay in the field is to go into industry. I'm just starting to get a sense of what kinds of jobs are available, but they range from working for Big Pharma and established biotech companies like Genentech, to smaller startups, to founding your own thing and looking for VC funding directly. The work you're doing still looks like biology at those places.

There are also hybrid, semi-academic research labs like the Broad Institute and their ilk. A number of places in the Bay Area are pushing in that direction. They often center around cutting-edge technologies, where there's value in having a lot of people do this stuff in a more stable way than academia affords. Because whereas in academia you need to constantly be doing the new thing, and in business you need to constantly be meeting the bottom line, these hybrid labs let you work on a well-established research area without being beholden to the bottom line.

These kinds of institutes can develop expertise and partnerships in a way that still feels academically rigorous, but isn't tied to whether we make quota next month, and at the same time isn't dependent on the novelty of the work. They just need to incrementally improve so that the labs or businesses they partner with and sell services to can achieve a much higher throughput, and do so much more cheaply than they could otherwise.

They can also provide an avenue for private-sector companies to come in and say, "Hey, we want to learn how to do this technique that was published two years ago. We're just reading the paper and to reinvent it from scratch would be really challenging, and there's not an option to send someone to go work in a research lab at MIT to learn it, so can we hire you to help us implement it?"

So these independent labs can form these elaborate partnerships, and some pretty exciting work happens in them. In terms of salary, they're never going to be fully competitive with Genentech or something, but they can often thread the needle in terms of exciting work and mostly competitive pay.

Then there are many largely tangentially disciplines you could pursue as well, like public policy or various forms of intellectual property law.

Institutional Ecosystems

Within academic research labs, how does the flow of money work? Where does funding come from, and who decides how it is spent?

I'm in an institution that is very heavily NIH-funded so I may have a skewed perspective on this, but my sense is that a lot of the funding for biology research in the country comes from the NIH. That shapes how people pitch their projects and what kinds of projects get funded.

Mechanically, the way it typically works is that there are grants — and I admit that I have only the loosest understanding of how those grants work — and there are different tiers of grants. I think the most fundamental one is called an R01.

If you are a principal investigator (PI) of a lab at a major institution, you submit an application making a case for why the work you're doing is important, why you're the right person to do it, why the place you're doing that work is the right place for it to be done, and why you'd be doing even more good things if you had more money. As part of your application, you also have to lay out your lab costs: do you need a giant, thirty-thousand-dollar centrifuge, or a PCR machine, or to buy time at a sequencing facility? And then you also have to account for the cost of the people that you have working in the lab.

> *"Fundamentally, there's a PI and then there's everybody else. After the PI the hierarchy is pretty flat, but the PI has absolute authority over the lab; it's basically a dictatorship."*

If you're awarded a grant, then the NIH effectively earmarks this big lump sum for you. They dole it out in increments and, depending on the grant, it can last several years before you have to renew. But it's much easier to maintain and renew an R01 than it is to get a new one. That's part of why it's hard to make it in academic biology: not only is there a finite amount of money, but if you're a new researcher trying to get your hands on an R01, it's this zero-sum game where incumbents have a leg up.

What is the breakdown and hierarchy of the roles in a lab like that?

Fundamentally, there's a PI and then there's everybody else. After the PI the hierarchy is pretty flat, but the PI has absolute authority over the lab; it's basically a dictatorship. The NIH gives them money, and it's like a VC giving money to a CEO: if you're a researcher in the lab, you're an employee.

As a PI, you basically get a grant based on what you said you were going to do, or had already done at the start of the grant. Then you give updates over the course of the grant, saying that you're doing useful stuff. It doesn't have to be the same stuff you said you were going to do, as long as it's useful. And then when you need to apply to re-up the grant four years down the road, there's a renegotiation and it's important that you've done impactful scientific things in the interim. In theory, it's possible to lose your grant at that time, but in between those goalposts you're otherwise pretty safe. If you were completely delinquent, they might strip your grant midstream, but it's really unlikely. It's like how you may not get elected for a second term as president, but you're less likely to get outright impeached.

Postdocs are senior to graduate students, insofar as postdocs already have their PhDs. Often, they are pursuing and acquiring their own sources of funding through lower-level grants that the NIH gives out to support nascent researchers to foster the next generation.

Graduate students are also pursuing different sources of money. The more money you bring in, the more independence you have because, at some level, the PI can tell me what to do with their money, whereas if I have my own money, I can spend it the way I want. But most of the money is coming from the top-level grants and from the PI.

In terms of autonomy and responsibilities within the lab, generally speaking, postdocs have more experience and are planning more of their own projects. They have a lot more flexibility to pursue their own ideas. The graduate students are beholden to their thesis committee, and also to what the lab can support. So if I have a fundamentally new idea as a graduate student and I can get my PI on board, that's fine, but it's going to be hard for me to get permission to access the slush funds to explore unilaterally.

There's another level below graduate students—well, parallel to or below depending on the culture of the lab—which is technicians. Their fundamental job is to analyze the science you've done, or perform an established scientific protocol at the bench. You know, turn the crank. If we need to process plasmid DNA extracted from bacteria four hundred times over the next three weeks, that's the technician's job, so they just do a lot of that. It's not specifically engaged with the intellectual-pursuit portion of the work. In a lab that doesn't have a lot of money, that may be a graduate student or postdoc's job. But if you can afford a tech, you'd rather have them doing that work.

You also have people who are in a nebulous region above postdocs: research scientists who are not PIs, but who help do research. They're not a canonical part of the lab. But there are labs, for example, where both members of a romantic partnership are scientists, and rather than founding two separate labs, they just decide to work together—so one of them becomes a researcher in the lab, getting paid by the research grants of their partner who is the PI. I know someone like this who is an

extremely senior researcher who probably could have founded
their own lab, but they'd rather work with their partner than
prioritize that prestige. I get the appeal.

How Not to Wipe Out the Human Race

**What is the conversation in your field around the ethics of
what you're doing? Is that something that explicitly comes
up?**

It varies based on what you're doing. There are some areas
where, in order to understand what's happening, you have to
work with humans. The biology of *Bacillus subtilis* and the biology of humans are extremely different, so I can do basic science
on bacteria all I want, but we don't really know how this thing
works in a human being until we've done it in human beings.

The cost of actually doing that is very high, in every possible sense. So people slowly work their way up, starting with
experiments on eukaryotic models like yeast. Then if it looks
promising, you move to mice, and then if it looks promising
with mice, you try monkeys, and then you do phase one clin-
ical trials with humans. All of this is far outside my ken, but
the point is that the more you move up this ladder, the more
you have to justify that what you're doing will ultimately have
positive health outcomes for human beings.

If I'm doing studies on mice, that usually means I have to kill
them after a month. There are whole review boards just making sure that if we're doing something harmful, it's within
reason. As soon as you climb north of yeast, there are more
controls — reviews to make sure you have thought about what
you're doing and convinced a panel of ethical experts that it's
responsible, and is aimed at the right ends. There are medical

ethicists, bioethicists, and, if you want money from the NIH, federal review boards that evaluate your research before you can get funding.

Even working with bacteria, there's still a certain amount of oversight, but it has less to do with ethics. If the things I wanted to do to bacteria were motivated purely by the desire to see them suffer, nobody would need to thwart me by founding People for the Ethical Treatment of Bacteria—I'd just never get funding, because who the hell would pay for that?

So ethical oversight is part of receiving funding.

Yeah, it's part of receiving funding and continuing to get funding and just having permission to do what you're doing.

Safety is the bigger issue for us when it comes to working with bacteria. It's not like anyone cares about how badly you treat them—our worry is that the *bacteria* might actually come kill *us*. So as long as you're doing a *thorough* job of killing them, review boards are fine with it. It's not about how badly they suffer on the way down the sink; it's about how much bleach did you pour down the sink with them.

All of those things involve different kinds of ethical questions and safety criteria, and there are good and bad systems in place. It's very good that we have protections for safety and for the ethical treatment of organisms, but there's a lot of random hit-or-miss stuff that winds up percolating through OSHA (the Occupational Safety and Health Administration).

Like what?

For example, there are chemicals that intercalate into DNA. When you're trying to identify things about DNA, often you want to "stain" it; you want to put something in it that becomes visible when you illuminate it with fluorescent light. And so you use these intercalating chemicals that get woven into the bands of the DNA, like hairs threaded through a comb. That's great, but it turns out that the presence of those chemicals makes the process of replicating the DNA less robust; as the DNA replicate themselves, errors get introduced. These cells that have random mutations might be cells in your body. Mostly, DNA errors just make cells less functional and they die, but occasionally these "mutations" interact with the cells in a way that makes them cancerous.

So scientists have identified that *sometimes* intercalating chemicals can be carcinogenic under certain circumstances, though we don't understand that very well. In the few cases where we've identified that a chemical might be a carcinogen, we've created these massive, Old Testament-style fences around it, like, "Thou shalt not come within a nine mile radius of anything labeled ethidium bromide, one of those fluorescent tagging chemicals, unless X and Y and Z and W conditions are met in the lab."

And then there are something like nine other chemicals like SYBR Gold and SYBR Safe, which are brand names for things that perform the exact same function as ethidium bromide, and I can keep those in a little cardboard box in my desk drawer, and nobody gives a shit because it's not ethidium bromide. Meanwhile, we're giving ethidium bromide to cows as an antibiotic because it's more harmful to bacteria than it is to eukaryotes. I could probably go drink it and be fine.

So I have to jump through these elaborate hoops to use this thing that probably isn't going to kill me, whereas there's almost no regulation at all of this other thing that's probably just as toxic. It's really arbitrary.

To take another example, there are probably ways in which, as an ethical researcher, you might be personally inclined to treat mice well even though there's no actual stipulation for how you make their lives more or less comfortable in the lab—and then other ways in which their lives are incredibly finely regulated, and the mouse can't actually tell the difference between a Level Eight and a Level Nine mattress in terms of how comfortably they are sleeping, but you need to always pick one or the other because there's a regulation about mouse mattresses.

I would guess that there's probably some deep bureaucratic misfirings that go into how this stuff gets regulated, because it's hard for laws to keep up with the pace of what the science is doing. It's this constantly moving target.

Regulations aside, how do scientists themselves talk about these issues?

While I'm sure there are callous assholes out there, for the most part biologists care about doing things in a way that is ethical, and healthy, and lets you sleep at night—even if it's just about cover-your-ass self-preservation. In my work with bacteria, theoretically we might do the wrong things and produce organisms that are harmful to humans, but we care a lot about not screwing up. I didn't sign up for this to wipe out the human race.

Scientists are particularly careful when technologies become clearly relevant to human health, like with CRISPR. There are several consortia, globally and nationally, where scientists are sitting down and saying, "Hey, we need to think about this, and here are our current thoughts. We need very clear boundaries and even moratoria in some cases to forestall negative outcomes."

The unfortunate reality is that those initiatives are like test ban treaties, in the sense that they work insofar as people are paying attention to you, but you don't have supreme dictatorial control over the world. Plus, there are guidelines, but the interpretation and implementation of those guidelines isn't uniform. And in a cutting-edge field of research like CRISPR, there are people out on the fringes in countries with less oversight who think, "That's cool, but if I'm the first one to do this research, I will be taken seriously. So I can either follow your regulations and nobody will ever care about the work I'm doing, or do this thing right now, in which case you guys will pay attention to me." So there's an obvious incentive to behave badly.

> *"I would guess that there's probably some deep bureaucratic misfirings of how that stuff gets regulated, because it's really hard for laws to keep up with the pace of what science is doing."*

In your experience, how do these conversations around ethics or oversight differ in the private sector?

The two worlds differ substantially when it comes to the role of ethics in funding. When you're trying to get VC funding, questions about ethics and safety are probably not the first ones that come up. But if you're trying to get money from the NIH, you've had to address ethics in a form as part of the application process, and they're going to keep checking in on it.

In tech, a company's vibe early on might be, "Let's just make a cool product for people." Then, as soon as you have accounts and authentication and enough money flowing through your system that there can be meaningful fraud, you start worrying about privacy and security. When you're starting to worry about those problems, you're already succeeding.

In biology, you have to worry about those problems *before* you even have a chance to scratch the surface of your research, because you're often trying to work on some disease that's killing people. We should worry deeply about whether our work is effective. If it's not, the best-worst case is that we're wasting money and implicitly killing people who have this rare illness because we're not finding a cure fast enough. The worst-worst case is that we're creating a thing that will *actively* kill people.

The way you get funding for VC-driven startups isn't necessarily by being super reliable and thoughtful and ethical. And I don't think there's a better example than Theranos. What Theranos was trying to do was at least noble in spirit. The people who funded them were a bunch of Silicon Valley VC firms, based on Beltway DC people vouching for them. Actual biologists were like, "That doesn't make sense." But it didn't matter.

I'm not saying all biotech startups are terrible—I think there's a lot of great work being done—but it's very hard to evaluate right now. I won't name names, but I can think of a couple of other companies where I'm like, well, that's either gonna be a scandal or a very quiet fizzle sometime soon. I've had good friends go to work for companies who are like, "What the fuck are we doing in my company? We've literally been injecting mice with saline solution and then looking at RNA sequencing to see what's happening. As far as I can tell, we are doing a deep dive on the placebo effect."

I don't know what's happening there. That's not even a joke. That's a real example from people I know. ᨒ

A Repair Manual for Spaceship Earth

by Alyssa Battistoni

A failed experiment in the Arizona desert holds valuable lessons for earthly survival.

On September 26, 1991, surrounded by the cameras of the world media, eight people dressed in bright red jumpsuits sealed themselves inside a three-acre steel-and-glass dome in the Arizona desert filled with over three thousand species of animals and plants. They planned to remain inside for two full years, aiming to show that the structure—known as Biosphere 2—was capable of sustaining life while completely sealed off from Biosphere 1, also known as Earth.

Amidst Biosphere 2's seven biomes—desert, rainforest, savannah, marsh, ocean, city, farm—the Biospherians would grow their own food and conduct research on the workings of the closed system. They would rely on the plants and animals they lived alongside to produce oxygen, absorb carbon dioxide, fertilize the soil, and consume waste. Lessons from the experiment were expected to advance the prospects of human life in space and on other planets.

Biosphere 2 may seem to be little more than a bizarre episode in the annals of extravagant scientific undertakings. But we should take its history seriously as we think about the future of life on Biosphere 1, which today appears fairly dire.

In the summer of 2019, Greenland's ice sheet lost nearly 200 billion tons of ice—three times the regular amount—while peat fires blazed across the Arctic. Two hundred reindeer starved to death in Norway while nearly two hundred gray whales have washed up dead on the western shores of North America since January. And that's just in the past year in one part of the world. The UN warns that a million more species are threatened with extinction in the next few decades as a result not only of climate change but overfishing, deforestation, and unsustainable agricultural patterns.

You can understand why someone might want to build another world.

"Is there a substitute for the work of nature—the work on which all other work depends?"

Everything You Can Do, I Can Do Better

These incidents suggest that the ecological systems we usually take for granted—sometimes referred to as Earth's "life-support systems"—are starting to break down. Healthy ecosystems are generally self-renewing. They operate without humans having to do much. They are, in a sense, already automated, at least from our perspective. But climate change and other ecological pressures are interrupting their normal function.

Climate change has caused more rain in the Arctic, which then freezes into ice, making it hard for reindeer and herbivores to find food. Melting Arctic sea ice, meanwhile, may have reduced the amount of algae in Arctic seas, which feeds the amphipods that feed the whales. The prospect of further disruption raises a question: Is there something we could do to fill in the gaps? Is there a substitute for the work of nature—the work on which all other work depends?

The fear that we're running down nature's abundance has a long history, of course. For economics—which is, after all, the study of scarcity—it's nothing to worry about. Conventional economic theory holds that natural resource scarcity can be solved through substitution. When resources become scarce, they become more expensive, which leads people to use them more efficiently or to use other, more plentiful materials in their stead.

As the growth economist Robert Solow once quipped, "The world has been exhausting its exhaustible resources since the first cave-man chipped a flint." In the eighteenth century, Thomas Malthus had worried about the ability to produce enough food; in the nineteenth, William Jevons had worried that the world would run out of coal. But new techniques, technologies, and resources overcame existing limits: petroleum replaced whale oil; steam engines replaced horses.

In the twentieth century, our capacity to create substitutes grew immensely. Many synthetic products were invented to take the place of natural ones. Declining soil nutrients could be replaced with artificial fertilizer; aluminum could replace copper; plastic could replace just about everything—wood, stone, metal, glass.

Nuclear power appeared poised to offer cheap, near-limitless energy supplies in place of fossil fuels extracted from the earth.

These advances gave rise to a way of thinking that we might call "substitution optimism": the belief that humans can find substitutes for anything that nature does. But substitution optimists tend to neglect two problems. First, the development of substitutes assumes that the price of scarce goods will rise. What about scarce goods that don't have a price? In particular, what about the services freely provided by nature? The services of atmospheric cycles and pollution-absorbing forests cost nothing—which mean that as they grow scarcer they do not get more expensive, and do not spur the development of technological replacements. Today, those resources—what we might think of as the earth's *reproductive* rather than productive functions—are the ones most under threat. Like human reproductive work, they operate in the background of economic production, providing the basic functions necessary for life.

But it's also an open question as to whether those kinds of resources actually have substitutes. Plastic chairs can substitute for wooden ones, or plastic bags for paper—but can you build a substitute for an entire forest? Can human technologies or human labor substitute for the nonhuman work done by other organisms? Or are there certain kinds of work that only nature can do?

Today's substitution optimists remain bullish. A group called the Ecomodernists, whose members include famed cultural entrepreneur Stewart Brand, geoengineering researcher David Keith, and Breakthrough Institute founders Ted Nordhaus and Michael Shellenberger, has taken up Brand's famous injunction from the *Whole Earth Catalog*: "We are as gods and might as well get good at it." Despite the signs of destruction all around, they assure us that human powers can yet be channeled to produce a "good Anthropocene."

In their view, resource scarcity isn't a problem: the Ecomodernist Manifesto of 2015 declares that "substitutes for other material inputs to human well-being can easily be found if those inputs become scarce or expensive." There are no real limits to growth: the sun provides more energy than we can hope to use, and any other given physical resource can be replaced with something else. That implicitly includes nature's reproductive functions. Carbon capture-and-storage technologies can replace a forest's capacity to absorb carbon. Injecting aerosols into the sky to make clouds more reflective mimics volcanic eruptions that spew sulfur into the atmosphere, helping to cool the earth.

> *"Plastic chairs can substitute for wooden ones, or plastic bags for paper — but can you build a substitute for an entire forest?"*

If Ecomodernists represent one extreme, the other end of the spectrum is occupied by those who spurn any kind of substitution. "Deep ecologists" see all of nature as intrinsically valuable: it's simply impossible to substitute for the unique and irreplaceable value of any given organism. For other ecologically minded thinkers, including proponents of "degrowth," the prospect of substituting technology for complex natural processes that we don't even fully understand is a typical demonstration of human arrogance, one that's certain to result in unintended consequences. In this view, technology is synonymous with the "techno-fix," a futile attempt to avoid deeper social and economic change through innovation.

Neither of these positions is satisfying. It's true that the Ecomodernists are wildly optimistic about human capacities and willfully obtuse about their limits. But it's not enough to smugly tut-tut at human hubris while the planet burns. Given how quickly the effects of climate change are materializing, even drastic decarbonization is unlikely to stop more mass die-offs and other forms of ecosystem dysfunction. We should hope that at least some ecosystem activities have substitutes, even if they can't be perfect ones.

"The question posed by Biosphere 2 was whether the entire Earth was substitutable"

In her 1970 book *The Dialectic of Sex*, best known for advocating artificial wombs as a substitute for biological ones, the feminist thinker Shulamith Firestone also called for a revolutionary ecological program. Such a program should seek to seize "control of the new technology for human purposes, the establishment of a new equilibrium between man and the artificial environment he is creating, to replace the destroyed 'natural' balance," she wrote.

Firestone, to be sure, had too much confidence in the possibility of liberation through technology, and too much fondness for the project of dominating nature. We have yet to automate human reproduction, and we're similarly unlikely to exert total technological control over Earth's reproductive functions. But we should nevertheless take seriously Firestone's impulse to see technology as part of the project of making a liberatory and livable planet rather than aiming for an impossible return to a natural balance that's everywhere in shambles and that in any case was never so harmonious as we imagine. We don't have to build the equivalent of an artificial womb for the entire Earth. But we should think about how to use our technologies for purposes both human and nonhuman, in a world where nature and human artifice are now so thoroughly entangled as to be inseparable.

The story of Biosphere 2 offers a way of thinking through what that might look like—both its possibilities and limitations.

The Garden and the Aircraft Carrier

The question posed by Biosphere 2 was whether the entire Earth was substitutable. The biosphere, a concept first developed by the Soviet scientist Vladimir Vernadsky, refers to the thin layer of the planet capable of supporting life. Biosphere 2 sought to replicate those life-support systems. The countercultural figure behind it, John Allen—an eccentric visionary with a degree in engineering, an MBA from Harvard Business School, and an enthusiasm for theater, poetry, and alternative living—saw the project as simultaneously an experiment with utopia and a backstop against dystopia.

"We poise ready now [sic] not only to cooperate consciously and creatively with the evolutionary potential of our present biosphere," he proclaimed, "but also to assist in its mitosis into other biospheres freeing our earth-life to participate in the full destiny of the cosmos itself, both by giving the possibility to voyage and live throughout space." Biosphere 2 would help bring about a new age of space exploration, Allen believed, by developing a way to sustain human life in hostile environments. It would also help protect human life against looming existential threats on Earth: Biosphere 2 would be a prototype of what Allen called "Refugia," self-contained living spaces that could serve as "insurance" against the calamity of a nuclear winter.

Before Biosphere 2, Allen had managed an intentional community outside of Santa Fe organized along ecological principles. There he met Edward Bass, an eccentric child of the Bass oil dynasty who got into ecology in the 1970s. By the 1990s, Bass was the largest private sponsor of environmental research in the United States. He funded an Institute for Biospheric Studies at Yale and smaller research projects around the world. He also poured an estimated $150 million into Biosphere 2 through his company Space Biosphere Ventures.

Bass saw these contributions as investments in projects that would one day become profitable: in the early days of the biotech boom in the 1990s, it seemed eminently plausible that "ecotech" would take off too. The initial income from Biosphere 2 was to come from tourism—they charged people $12.95 to visit, and half a million people did—which would support the longer-term development of technologies that Bass expected "would have a very significant commercial application."

Biosphere 2 was, as one of its inhabitants called it, "the garden of Eden on top of an aircraft carrier." Underneath the rainforest and desert landscapes was a massive technosphere, comprising three acres of electrical, mechanical, and plumbing systems. The technology was intended, as a Biospherian put it, "to replicate many of Earth's free services"—the reproductive functions. On Earth, various planetary processes keep air and water moving, and nutrients and waste along with them. In Biosphere 2, that work was mechanized.

Some machines treated wastewater and desalinated water from the miniature ocean to make rain. Others created breezes, mists, and ocean waves. Giant air handlers heated and cooled the air, running off generators powered by natural gas and diesel. To prevent the domes from exploding as the air inside warmed and expanded in the heat of the desert sun, a giant pair of rubber "lungs" acted as a safety valve.

These innovations drew on two decades of research on how to keep humans alive in space, most notably a proposal by the twentieth-century ecologists Eugene and Howard Odum to build "bioregenerative life support systems" in spacecraft. Most spaceships were all technosphere: everything humans needed to survive was provided by a machine. The Odums' idea was to replace some of those technological functions with organisms that could perform the necessary functions of oxygen generation, waste removal, and food production. Bioregenerative systems would bring down the cost of space travel, reduce the need for astronauts' labor, and make it possible for astronauts to live longer in space without continually receiving supplies from Earth.

Biosphere 2 would be the most ambitious embodiment of these ideas to date. When it opened, it was heralded by many in the press as a marvel of both technological and ecological engineering — *Discover* magazine called it "the most exciting scientific project" since the moonshot — even as many scientists looked on skeptically. The crew that entered Biosphere 2 in 1991 was to be the first of many: Allen imagined that new crews would enter every two years for an entire century, each building on the knowledge of those who had gone before. Cumulatively, they would inaugurate a new era in the understanding of life on Earth — and the possibilities of life beyond.

The Bubble Bursts

Things didn't go as planned. If the technology of the "aircraft carrier" was cutting-edge, what lay above wasn't exactly the garden of Eden. Once the Biospherians were sealed inside, everything began to go drastically wrong. Though they had attempted to carefully calibrate the equilibrium of the internal ecosystem, an artificial balance was hard to strike.

Cramming seven biomes into just three acres led to some unexpected effects: the desert picked up condensation from the forest and became more like a shrubland. Nor had Biosphere 2 managed to replicate all of Earth's services: many trees in the rainforest and the savannah that would usually grow "tension wood" in response to winds failed to do so in the calm Biosphere, leaving them weaker. Most troublingly, the Biosphere began to lose a huge amount of oxygen, while carbon dioxide and nitrous oxide levels rose dangerously. The Biospherians tried to sequester carbon by growing plants, and stopped tilling the soil to prevent carbon stored in the soil from being released into the air. But they couldn't figure out how to actually stop carbon from accumulating.

It turned out that microbes in the soil were producing carbon dioxide faster than plants were producing oxygen, while the structure's concrete foundations were absorbing a surprising amount of oxygen. Some speculated that the El Niño event of 1991–1992 also contributed to more cloud cover than usual, decreasing the amount of sunlight for plants to photosynthesize. Some of the vines that the Biospherians had planted to absorb carbon started to overtake food crops, requiring intensive weeding. Algae consumed the ocean, requiring Biospherians to clear it away by hand so the coral reefs below could receive sunlight.

The complications multiplied. An estimated 30 percent of the 3,800 enclosed species died off, including all pollinators. The Biosphere was overrun by ants and cockroaches, stowaways inside the sealed system that soon outcompeted and outlasted other insects. By the time oxygen levels had dropped from 20.9 percent to 14.2 percent—the equivalent of living at an elevation of 15,000 feet—it became difficult to breathe, at which point the Biospherians broke the closed system to pump in oxygen and keep the crew alive.

"Once the Biospherians were sealed inside, everything began to go drastically wrong."

The first crew left Biosphere 2 in September 1993, on schedule and underweight. The second crew of eight entered in March 1994. But by then Biosphere 2 was looking more like a boondoggle than a breakthrough: it cost a great deal to maintain, and seemed unlikely to develop a commercially viable product anytime soon. Bass began feuding with Allen and then fired most of the Biosphere 2 staff while the second crew was still inside. He hired Steve Bannon, at the time an investment banker experienced with company takeovers, to manage Space Biosphere Ventures's financial affairs. Amidst the turmoil, the second crew left the structure six months later. They would be the last people ever to live inside Biosphere 2.

Bannon brokered a deal with Columbia University, which agreed to take the facility over. (Columbia eventually gave it up, citing exorbitant expenses; in 2005, Bass gifted Biosphere 2 to the University of Arizona, which now runs it as a research facility.) Space Biosphere Ventures ended up facing multiple lawsuits. Though the irresistible spectacle of Biosphere 2 had made it a media darling at the outset, as the project faltered it was decried as a stunt, a hoax, a fraud. *The Village Voice* described Biosphere

2 as the product of "an authoritarian—and decidedly non-scientific—personality cult." The fact that the closed system had been breached to restore oxygen levels rendered the scientific value of the grand experiment dubious. Academic scientists, vindicated by the downfall of a flashy for-profit interloper, set about diagnosing the causes of the disaster.

The most obvious lesson was that replicating the reproductive functions of Earth was much more complicated than anyone had imagined. As a pair of Columbia researchers wrote in an assessment shortly after Columbia took over the facility, "isolating small pieces of large biomes and juxtaposing them in an artificial enclosure changed their functioning and interactions rather than creating a small working Earth as originally intended."

"If we're going to do more than mourn or panic, we have to take the idea of substitution seriously."

You could not simply treat ecosystems as mechanical pieces to be assembled and slotted in and out. Ecologists didn't even know what all the pieces of an ecosystem were, let alone how exactly they worked together. "At present there is no demonstrated alternative to maintaining the viability of Earth," they concluded. "No one yet knows how to engineer systems that provide humans with the life-supporting services that natural ecosystems provide for free." Ecosystems did not appear to be very substitutable at all.

Jury-rigging Spaceship Earth

The story of Biosphere 2 seems to prove the substitution skeptics right: we can't replace ecosystems and we shouldn't try. But I can't muster much schadenfreude about the failures of Biosphere 2. After all, the misfortunes of the Biospherians look worrisomely like our own.

Even if we manage to stop climate change from reaching truly cataclysmic levels, rising temperatures will transform Earth's systems in ways that will make it difficult for many species to survive. Under the circumstances, pious affirmations of ecological holism can quickly tilt into premonitions of doom: if ecosystems are beyond human understanding and entirely irreplaceable, collapse is only a matter of time.

The biologists Paul and Anne Ehrlich once compared species to rivets on an airplane wing. If you were in a plane and looked out the window and saw a rivet fall off the wing, you might be a little concerned but not too worried — after all, the wing has thousands of rivets, enough to make any single one redundant. But if lots of rivets started popping off, you would probably start to freak out. Similarly, losing a species or two might be worrisome but not a sign of doomsday. Losing a lot of species, however, suggests that Spaceship Earth might be in trouble.

We are going to lose more rivets. I hope we can jury-rig something to keep the plane in the air. Our lack of control over the biosphere is genuinely terrifying. But if we're going to do more than mourn or panic, we have to take the idea of substitution seriously.

Instead of treating substitution as the frictionless replacement of one kind of thing for another, as if matter were totally commensurable, however, we could recognize that substitutes might be rough around the edges but can nevertheless help prevent total breakdown. In some cases, substitution might mean using one kind of organism in place of another; in others, it might mean substituting human labor or technology for the work of nature.

The idea that one species could do the same work as another was one of Charles Darwin's great insights. As the environmental historian Donald Worster relates, when Darwin went to the Galápagos, he noticed that giant tortoises did the grazing work that bison did in North America. Different creatures, that is, held "the same place in Nature"; they could fill the same "office" within an ecosystem—what ecologists would eventually come to call a niche. Different organisms would do the job differently, of course: a tortoise would have different predators and reproduction patterns than a bison, even if they both grazed. But ecosystems weren't fixed, timeless orders wherein each organism performed its appointed role for eternity. They were struggles to stay ahead of the competition or be replaced by something else.

This principle animated the selection of species in Biosphere 2. The thousands of species sealed in the dome were chosen not to faithfully replicate the exact relationships of existing ecosystems, but to provide particular functions: to serve as pollinators, to supply food crops, to recycle air, to decompose waste. If a particular species didn't fit the practical needs of the Biosphere, it was replaced with another.

Of course, human efforts to achieve particular ends by introducing new species don't always go well. The genre of stories about "invasive species" is one of the most reliable sources of cautionary tales about unintended consequences of human meddling in nature. The East Asian vine kudzu, for example, was widely planted in Southern states to help address soil erosion in the aftermath of the Dust Bowl; it became the weed that ate the South.

But not every story of introduced species is a warning. Rewilding projects, for example, attempt to restore land domesticated and cultivated by humans into ecosystems operating without human presence. This usually means reintroducing species that have been driven out of their former habitats or killed off by human settlement. In some cases, species have gone fully extinct, yet some other kind of creature may be able to do the same work.

In the rewilded nature reserve of Oostvaardersplassen in the Netherlands, for example, the roles of extinct aurochs and tarpans—wild ponies and cattle—are filled by physiologically similar breeds that eat the same grasses and have similar roaming patterns. In the American Great Plains, meanwhile, sustainable cattle raising practices have tried to replicate the grazing patterns of bison. Since cattle tend to roam less extensively, doing so requires more intensive human labor to direct herds.

This raises an important point: natural systems that now operate automatically may require *more* human labor to function as nonhuman species disappear. Life in Biosphere 2 was a lot of work. As one participant later recalled, "Farming took up 25 percent of our waking time, research and maintenance 20 percent, writing reports 19 percent, cooking 12 percent, biome management 11 percent, animal husbandry 9 percent." As Biosphere 2's nonhuman life support systems started to falter, its human inhabitants had to work harder to keep them functioning, from chasing pests that had no predators to pollinating plants when the bees died off. As scientists observed in the aftermath, "Biospherians, despite annual energy inputs costing about $1 million, had to make enormous, often heroic, personal efforts to maintain ecosystem services that most people take for granted."

Of course, Biosphere 2 didn't just rely on human labor to help nature function. Its vast technosphere was built expressly to fill in for Earth systems like ocean currents and water cycles. Biosphere 1 now has a rather substantial technosphere of its own, currently constituting 30 trillion tons of human artifacts, from computers to undersea cables to houses to lightbulbs. Most of it supports human life and pursuits. But some of it could also be put to use tending to the biosphere, as in Biosphere 2.

The garden of Eden and the aircraft carrier aren't our only options. Most of our world is some combination of the two. Using technology to support ecological functions doesn't have to involve building a giant array of machinery to replace Earth systems or trying to technologically manipulate the entire atmosphere, a la the Ecomodernists. But nor should it mean attempting to remove human activity and artifacts from ecosystems altogether. As Donna Haraway reminds us, "There is no Eden under glass." Technology can play an important role in actively maintaining ecosystems rather than replacing them wholesale, in conjunction with human labor.

Some of this work is already happening. Drones are being used to reseed land for restorative purposes, effectively performing the work of birds while reducing human presence in remote areas. In the Great Barrier Reef, a robotic vessel protects indigenous coral species by killing the crown-of-thorns starfish that is suffocating the reef.

Paradoxically, these unmistakably human interventions often occur in the absence of actual humans. Robots can offer ways to preserve nonhuman ecosystems without more direct forms of human intrusion. They aren't total replacements for organisms, of course. A drone can drop seeds but can't lay eggs; a robot fish can kill starfish but can't grow new coral. Indeed, none of the options available to us — nonhuman proxies, technological tools, human labor — is a perfect substitute for what they replace, and none ever will be. At best, they can provide rough approximations of certain functions. But these jury-rigged rivets might be our best hope for making a future on a damaged planet.

"The garden of Eden and the aircraft carrier aren't our only options. Most of our world is some combination of the two."

How Are You Going to Pay for It?

Nature, Raymond Williams once said, is the most complex word in the English language. But I've come to think that "natural" mostly means "freely given." Nature offers the "free services" on which human life depends. More generally, nature describes what we take for granted, what we expect to happen of its own accord. From "natural birth" to "natural beauty," nature hides a lot of work done behind the scenes. As the scholar Merve Emre reminds us, "all reproduction, even reproduction that appears 'natural,' is assisted." Emre is concerned with human reproduction, but it holds just as true for the reproduction of nature itself.

We can no longer take the reproduction of our world for granted, or assume that the work of nature will take place automatically. Reproducing life on Earth will require a great deal more assistance from us, in our simultaneously extraordinary and limited capacities as a single species on a planet of millions. It will also require a great deal more recognition of the assistance provided by all those other species. What the feminist theorist Sophie Lewis calls "full surrogacy"—a call to distribute labor more broadly, to cultivate reciprocal practices of kinship and care—is as applicable to our nonhuman relationships as to our human ones.

While we may be able to perform some work on nature's behalf in order to stabilize our biosphere, however, the expense will be enormous. Indeed, the biggest barrier to developing substitutes for certain ecological services may turn out to be cost.

The ecologist John Avise observed that the true lesson of Biosphere 2 was an economic one. In the late twentieth century, economists had tried to estimate the value of Earth's freely provided services, but had usually stumbled over the technical difficulties of doing so. Biosphere 2 made it possible to construct "a more explicit ledger," Avise wrote. All told, it had cost over $150 million to keep eight humans alive for two years. As Avise pointed out, "if we were being charged, the total invoice for all Earthospherians would come to an astronomical three quintillion dollars for the current generation alone!" Replacing human labor with machines usually saves money. Replacing the work of nature with machines or human labor is the opposite: it makes what was free expensive.

This means that substitution is rarely economical. In China's Hanyuan County, for example, where pesticides have wiped out many bee colonies, human workers have subbed in, using feather dusters to pollinate pear trees by hand. But human pollination is only viable in Hanyuan because it's cheaper than renting beehives. In a system (capitalism) that aims to keep costs down above all else, the cost of human labor has to be approaching zero for it to compete with nature's gifts.

So as we ask who, or what, will do the work of nature, we should also ask another question: Who will pay for it? Earthly survival will require new ways of organizing not only our social and technological relationships, but our economic ones. As Biosphere 2 demonstrates, filling in for the work of nature is unlikely to be a profitable enterprise. Capitalism is unlikely to pay the extra costs. The question of what can replace it may be the biggest substitution problem of all. ∿

Alyssa Battistoni is a political theorist and postdoctoral fellow at Harvard University. She is the coauthor of *A Planet to Win: Why We Need a Green New Deal*.

This issue is dedicated to the memory of
Annie the greyhound, now and forever
the official doge of Logic.

March 30, 2009 – October 2, 2019

o hi

We're a small magazine, and we believe in paying writers.

Around half of our expenses go to commissioning fees. Writing is hard work, and we believe we can't have a better discourse around technology without compensating the people who are working to improve it.

But we can't do this without your help.

Subscribing is one way to support this project. Making a tax-deductible donation is another.

Logic Magazine is published by the Logic Foundation, a California nonprofit public benefit corporation with 501(c)(3) status.

Your contributions will enable us to keep building a project that's committed to recognizing and rewarding creative labor.

For more information about the Logic Foundation, including how to donate electronically or by check, visit **logicmag.io/donate**.

Other questions? Email **donate@logicmag.io**.

thxbye

LOGIC

upcoming

ISSUE 10: Security

Since the beginning, humans have made tools in order to protect ourselves and our loved ones. The simplest hut shelters you from at least some of the elements; a bow and arrow spares you from doing hand-to-hand combat with a sabertooth tiger; a writing implement lets you tell future generations how to do the same. And yet, even now, there are no guarantees. Internet transmissions are inherently leaky. Becoming "smart" can make the most banal object hackable: your toaster becomes part of a botnet; your baby's crib is a spy. Digital platforms have created new kinds of precarity, as they disrupt workplaces and algorithms handle scheduling and benefits. This issue will look at how we use technologies to stay safe — and the novel dangers that these same technologies create.

ISSUE 11: CARE

SUMMER 2020

Digital technologies tend to be depicted as steely or ethereal. A headless suit holds a giant computer chip. A bodiless hand holds a cell phone. A brain wired to a giant computer network bursts into rainbows of light. But technologies are not invulnerable — nor are the people who build and use them. The gig economy is not all Uber drivers — care workers are its fastest-growing demographic. This issue will look at technologies that are changing how we give and receive care — and the care that our machines themselves need.

subscribe @ https://logicmag.io